CYTOTOXINS AND IMMUNOTOXINS FOR CANCER THERAPY
Clinical Applications

EDITED BY

Koji Kawakami

Bharat B. Aggarwal

Raj K. Puri

CRC PRESS

Boca Raton   London   New York   Washington, D.C.

Library of Congress Cataloging-in-Publication Data

Cytotoxins and immunotoxins for cancer therapy : clinical applications/edited by
Koji Kawakami, Bharat B. Aggarwal, Raj K. Puri
    p. cm.
Includes bibliographical references and index.
ISBN 0-4152-6365-4 (alk. paper)
    1. Cancer — Chemotherapy. 2. Cytokines — Therapeutic use. 3. Antibody-toxin
    conjugates. I. Kawakami, Koji. II. Aggarwal, Bharat B., 1950– . III. Puri, Raj K.
    [DNLM: 1. Neoplasms — drug therapy. 2. Clinical Trials. 3. Cytotoxicity,
    Immunologic. 4. Cytotoxins — therapeutic use. 5. Immunotoxins — thera-
    peutic use. QZ 267 C9977 2004]
RC271.C95C985 2004
616.99'4061—dc22

                2004050334

Visit the CRC Press Web site at www.crcpress.com

© 2005 by CRC Press

No claim to original U.S. Government works
International Standard Book Number 0-4152-6365-4
Library of Congress Card Number 2004050334
Printed in the United States of America 1 2 3 4 5 6 7 8 9 0
Printed on acid-free paper

Preface

An overview of immunotoxins and cytotoxins in clinical trial

By utilizing the knowledge of unique or overexpressed cell-surface antigens or receptors on tumor cells as targets, a new form of cancer therapy has evolved over the last two decades. In particular, a variety of receptors for cellular growth factors and cytokines have been shown to be overexpressed on tumor cells, which may possibly serve as targets for cancer therapy. These include fibroblast growth factor receptors, insulin-like growth factor type I receptors, epidermal growth factor receptors, transferrin receptors, vascular endothelial growth factor receptors, transforming growth factor receptors, interleukin-4 receptors, and interleukin-13 receptors.

It has been reported that neutralizing monoclonal antibodies either alone or attached with radionuclide or antisense oligonucleotides to some of these receptors can inhibit the growth of tumor cells by evoking host immune responses. For direct tumor cell killing, antibodies or ligands have been chemically conjugated or genetically fused to plant or bacterial toxins, i.e., *Pseudomonas* exotoxin (derived from *Pseudomonas aeruginosa*), diphtheria toxin (derived from *Corynebacterium diphtheria*), ricin, and saporin (derived from *Saponaria officinalis*). These conjugated or fused chimeric proteins are termed immunotoxins or cytotoxins.

Such immunotoxins and cytotoxins have been tested extensively in hematologic or nonhematologic solid tumors *in vitro* and *in vivo* in animal models. Furthermore, some immunotoxins such as Tac(Fv)-PE38 (LMB-2), RFB4(dsFv)-PE38 (BL22), IL4(38–37)-PE38KDEL, IL13-PE38QQR, Tf-CRM107, $DT_{388}GMCSF$, and cBR96-doxorubicin (SGN-15) are being tested in clinical trials. In addition, two targeted cytotoxin and immunotoxin, $DAB_{389}IL-2$ (Ontak™) targeting IL-2 receptors and CD33-calicheamycin (Mylotarg™), have been licensed by the FDA for cutaneous T-cell lymphoma (CTCL) and relapsed acute myeloid leukemia (AML), respectively. Although a large number of articles have been published on these cytotoxins and immunotoxins, no focused issues containing comprehensive lists of articles for clinical use have been published.

In this book, we provide a list of chapters dealing with various cytotoxins and immunotoxins. A number of distinguished authors have summarized their work and discussed their studies from discovery to *in vitro* testing, to preclinical models and finally clinical trials.

In Part I of the book, *Pseudomonas* exotoxin (PE)-based immunotoxins have been described. Dr. Robert Kreitman summarized preclinical and clinical development of Anti-Tac(Fv)-PE38 (LMB-2) that targets CD25-positive leukemia and RFB4(dsFv)-PE38 (BL22) targeting CD22-positive B-cell malignancies (Chapter 1). Recombinant immunotoxins targeting CD25 and CD22 have been tested in patients and produced major responses in several chemoresistant hematologic malignancies, particularly hairy cell leukemia. This section also describes the up-to-date results from clinical trials with IL-4 cytotoxin IL4(38–37)-PE38KDEL for the treatment of recurrent glioblastoma (Chapter 2) and IL-13 cytotoxin IL13-PE38QQR for the treatment of human cancer (Chapter 3). Preliminary results from an extended phase I/II clinical trials in patients with recurrent glioma suggest that direct intratumoral convection-enhanced delivery (CED) of IL-4 cytotoxin can cause pronounced necrosis of recurrent glioma tumors without systemic toxicity (Chapter 2). Four phase I/II clinical trials in patients with GBM and advanced metastatic RCC have been initiated and some completed. These clinical studies have demonstrated that IL-13 cytotoxin could be safely administered intravenously at a dose of up to 2 $\mu g/kg$ q.o.d $\times$ 3 in RCC trial and intratumorally for up to 4 $\mu g/ml$ in GBM trial (Chapter 3).

In Part II, Drs. Roland Schnell and Andreas Engert summarized the role of ricin-based immunotoxins for the treatment of lymphoma (Chapter 4). A variety of ricin-based immunotoxins demonstrated biologic activity in several clinical trials in Hodgkin's and B-cell non-Hodgkin's lymphomas, even in heavily pretreated patients with bulky relapsed disease, with varied response rates from 6 to 23%.

In Part III, several DT-based immunotoxins have been discussed indicating significant interest in DT-based immunotoxins. Chapter 5 describes the preclinical development of $DT_{390}IL13$ and DTAT (diphtheria toxin amino terminal fragment of urokinase-type plasminogen activator) by Dr. Daniel Vallera's group. Efficacy and toxicity of these agents in the flank tumor experiments are summarized. Drs. Paul Shaughnessy and Charles LeMaistre summarized in detail the preclinical and clinical development of the first FDA-approved cytotoxin, $DAB_{389}IL-2$ (Ontak™), for the treatment of CTCL expressing CD25 (Chapter 6). $DAB_{389}IL-2$ has demonstrated acceptable toxicity and substantial clinical benefit to patients with advanced and refractory CTCL. Drs. Simon Long and Patrick Rossi described the preclinical development and clinical studies using Tf-CRM107 for the treatment of malignant brain tumor (Chapter 7). Phase I and II clinical trials with Tf-CRM107 have suggested that the drug is able to induce responses in a significant number of patients with high-grade gliomas that have failed conventional therapy. In Chapter 8, Dr. Art Frankel and his colleagues have summarized preclinical and clinical studies with $DT_{388}GMCSF$ for the treatment of AML. They demonstrated that $DT_{388}GMCSF$ have some clinical activity as 4 of 37 patients showed clinical remissions of their disease. In Chapter 9, Drs. Indira

Subramanian and Sundaram Ramakrishnan described the preclinical development of VEGF121–DT385 and VEGF165–DT385 immunotoxins. VEGF–toxin conjugates made either by chemical linkage or by genetic fusion are effective reagents capable of inhibiting angiogenesis and tumor growth in animal models.

In Part IV, two important chemically conjugated immunotoxins are discussed. Drs. Lisle Nabell, Clay Siegall, and Mansoor Saleh summarized multiple phase I and II clinical trials with cBR96-doxorubicin (monoclonal antibody to Lewis[y] antigen conjugated to doxorubicin termed SGN15). The clinical trials with SGN-15 as a single agent or in combination with docetaxel have shown some clinical activity in patients with colon and breast cancer. Finally, Drs. Charalambos Andreadis, Selina Luger, and Edward Stadtmauer have described anti-CD33-calicheamicin immunoconjugate (gemtuzumab ozogamicin, Mylotarg™) in detail in Chapter 11 for the treatment of AML. Marketing approval of this agent was granted on May 17, 2000, by the FDA under the Accelerated Approval regulations.

The editors had envisioned focusing on clinical applications of these cytotoxins and immunotoxins for cancer therapy and we are pleased to report that most contributors kindly agreed and abided by this request and provided comprehensive knowledge and clinical developmental approaches of cytotoxins and immunotoxins for cancer therapy. Although as described above, some agents have been licensed for the marketing, most of these cytotoxins and immunotoxins remain in phase I/II stages of clinical trial development except IL-13-PE, which is being tested in phase III clinical trial for the treatment of recurrent glioblastoma multiforme.

Because of the nature of the phase I/II clinical trials, no definite conclusions can be made for efficacy of these exciting agents. Nevertheless, these agents remain exciting possibilities for cancer therapies. Therefore, editors believe that the information summarized in this book will be extremely valuable to scientists and clinicians who are involved in the development of immunotoxins and cytotoxins for the treatment of cancer. Finally, information provided in this book would benefit interested parties for the development of new targeted agents for not only cancer therapy but also for other diseases. No official support or endorsement of this book by the Food and Drug Administration is intended or should be inferred.

About the Editors

Koji Kawakami, M.D., Ph.D., is a staff member of the Division of Cellular and Gene Therapies in the Office of Cellular, Tissue, and Gene Therapies, Center for Biologics Evaluation and Research (CBER), U.S. Food and Drug Administration (FDA), Bethesda, Maryland. After his training as a head and neck surgeon in Japan, Dr. Kawakami has joined CBER/FDA and conducted a number of research projects in cytokine immunobiology, translational research, cancer gene therapy, and targeted cancer therapy. Serving as a regulatory product reviewer specialized in tumor vaccine and cancer gene therapy at the FDA, Dr. Kawakami reviews investigational new drug applications (INDs) submitted from U.S. industries and academic and governmental institutes.

Dr. Bharat B. Aggarwal received his Ph.D. in biochemistry from the University of California, Berkeley; completed postdoctoral training at the University of California Medical Center, San Francisco; and worked at Genentech, where his team isolated TNF-α and TNF-β. Currently, Dr. Aggarwal is Professor and Chief of the Cytokine Research Section at the University of Texas M.D. Anderson Cancer Center in Houston. He holds the Ransom Horne Jr. Endowed Distinguished Professorship. Dr. Aggarwal has published over 300 original peer-reviewed articles, edited 6 books and filed/received over 30 patents. He serves on the editorial boards of *Journal of Biological Chemistry, Cancer Research, Clinical Cancer Research, Journal of Immunology,* and others. Since 2001, Dr. Aggarwal has been one of the most highly cited scientists by the ISI.

Raj K. Puri, M.D., Ph.D., is Chief of Laboratory of Molecular Tumor Biology, Division of Cellular and Gene Therapies in the Office of Cellular, Tissue, and Gene Therapies, Center for Biologics Evaluation and Research (CBER), U.S. Food and Drug Administration (FDA), Bethesda, Maryland. Dr. Puri is also a Co-Director of the CBER/National Cancer Institute Microarray Program. Dr. Puri directs basic and translational research in the field of tumor biology and molecular targeting of human cancer with novel therapeutic approaches. Two of the targeted agents, interleukin-4-*Pseudomonas* exotoxin and IL-13-*Pseudomonas* exotoxin, invented in Puri lab are in clinical trials in the U.S. and Europe for the treatment of malignant glioma. His recent interest is focused on the characterization of tumor vaccines, human stem cells and tissue engineered products by microarray technology. Dr. Puri evaluates and supervises investigational new drug applications and biological license applications for tumor vaccines and gene therapy products for cancer.

Contributors

Bharat B. Aggarwal
M.D. Anderson Cancer Center
Houston, Texas

Charalambos Andreadis
University of Pennsylvania Cancer
   Center
Philadelphia, Pennsylvania

Dongsun Cao
Wake Forest University School of
   Medicine
Winston-Salem, North Carolina

Kimberley A. Cohen
Wake Forest University School of
   Medicine
Winston-Salem, North Carolina

Andreas Engert
University of Cologne
Cologne, Germany

Arthur E. Frankel
Wake Forest University School of
   Medicine
Winston-Salem, North Carolina

Philip D. Hall
Medical University of South
   Carolina
Charleston, South Carolina

Walter A. Hall
University of Minnesota
Minneapolis, Minnesota

Syed R. Husain
U.S. Food and Drug Administration
Bethesda, Maryland

Bharat H. Joshi
U.S. Food and Drug Administration
Bethesda, Maryland

Koji Kawakami
U.S. Food and Drug Administration
Bethesda, Maryland

Mariko Kawakami
U.S. Food and Drug Administration
Bethesda, Maryland

Robert J. Kreitman
National Institutes of Health
Bethesda, Maryland

Charles F. LeMaistre
Texas Transplant Center
San Antonio, Texas

Pamela Leland
U.S. Food and Drug Administration
Bethesda, Maryland

Tie Fu Liu
Wake Forest University School of
   Medicine
Winston-Salem, North Carolina

Simon Long
KS Biomedix, Ltd.
Guildford, Surrey, England

Selina M. Luger
University of Pennsylvania Cancer
   Center
Philadelphia, Pennsylvania

Marlena Moors
Wake Forest University of School of
   Medicine
Winston-Salem, North Carolina

Lisle Nabell
University of Alabama at
   Birmingham
Birmingham, Alabama

Raj K. Puri
U.S. Food and Drug Administration
Bethesda, Maryland

Sundaram Ramakrishnan
University of Minnesota Medical
   Center
Minneapolis, Minnesota

Patrick Rossi
KS Biomedix, Ltd.
Guildford, Surrey, England

Edward Rustamzadeh
University of Minnesota
Minneapolis, Minnesota

Mansoor N. Saleh
University of Alabama at
   Birmingham
Birmingham, Alabama

Roland Schnell
University of Cologne
Cologne, Germany

Paul J. Shaughnessy
Texas Transplant Institute
San Antonio, Texas

Clay B. Siegall
Seattle Genetics, Inc.
Bothell, Washington

Edward A. Stadtmauer
University of Pennsylvania Cancer
   Center
Philadelphia, Pennsylvania

Indira V. Subramanian
University of Minnesota–Twin
   Cities
Minneapolis, Minnesota

Andrew M. Thorburn
Wake Forest University of School of
   Medicine
Winston-Salem, North Carolina

Deborah Todhunter
University of Minnesota Cancer
   Research Center
Minneapolis, Minnesota

Daniel Vallera
University of Minnesota Cancer
   Center
Minneapolis, Minnesota

Contents

part one

Pseudomonas *exotoxin-based immunotoxins/cytotoxins*

chapter one

Targeting of Pseudomonas *exotoxin-based immunotoxins to hematologic malignancies*

Robert J. Kreitman

Contents

0-4152-6365-4/05/$0.00+$1.50
© 2005 by CRC Press

3

The hematologic malignancies, leukemias, lymphomas, and Hodgkin's disease, are often resistant to standard treatments of chemotherapy and radiation either primarily or after one or several relapses. Immunotoxins have been constructed that contain a monoclonal antibody and truncated forms of *Pseudomonas* exotoxin. Recombinant immunotoxins can be produced in bacteria (*E. coli*) as ligand-toxin fusion proteins, and do not require chemical conjugation of the antibody and toxin. This is because one of the variable domains of the antibody (Fv) can be fused to the truncated toxin, and the other variable domain can connect to the first either through a peptide linker or an engineered disulfide bond. Recombinant immunotoxins targeting CD25 and CD22 have been tested in patients and have produced major responses in several chemoresistant hematologic malignancies, particularly hairy cell leukemia (HCL).

Introduction

Therapy of hematologic malignancies

It is predicted that 106,000 patients with hematologic malignancies will be diagnosed in the United States in 2003, and that there will be 47,000 deaths in this group of patients.[1] The majority of these patients will require systemic therapy for widespread disease; many of these patients experience failure of chemotherapy, either at diagnosis or after one or several relapses. Relapse is often diagnosed using immunologic techniques such as flow cytometry or immunohistochemistry, since tumor associated antigens on the surface of such cells remain despite the changes inside the cell that cause chemoresistance. To target such cells, a variety of surface-targeted monoclonal antibody therapies have been developed.

Immunotoxins in the hierarchy of surface-targeted therapies for hematologic malignancies

Today, the most widely used type of surface-targeted biologic therapy is unlabeled monoclonal antibodies (MAbs), such as rituximab[2] and alemtuzumab.[3] These MAbs, containing mostly human IgG sequences, are effective in up to half of patients via mechanisms of apoptosis induction, antibody dependent cellular cytotoxicity, and complement dependent cytotoxicity. Patients whose malignant cells are resistant to apoptosis, or patients who lack an adequate immune system, may be resistant to unlabeled MAbs. Surface-targeted therapy in such patients requires a more passive immunotherapeutic approach. To kill cells more directly, MAbs have been conjugated to radionuclides.

Examples of these agents include Bexxar[4] and Zevalin,[5] anti-CD20 MAbs radiolabeled with ^{131}I and ^{90}Y, respectively. Such agents can induce responses in lymphoma patients even after pretreatment with rituxan. However, the dose is usually limited by cumulative nonspecific uptake into normal marrow, most often causing thrombocytopenia. A third type of surface-targeted therapy is exemplified by the recently approved Gemtuzumab ozogamicin, a conjugate of anti-CD33 MAb and the small molecule cytotoxic drug calicheamicin.[6] This type of therapy in patients with relapsed/refractory acute myelogenous leukemia can induce major responses in about 30% without the toxicities of chemotherapy.[7] It has been shown that multidrug resistant cells are resistant.[8] To target the surface of chemoresistant hematologic tumor cells in immunocompromised patients, a fourth type of therapy has been used, involving MAbs connected to protein toxins.

Targeted toxin therapy

Protein toxins kill cells by inhibition of protein synthesis after a toxin fragment enters the cytosol of the cell. Plant toxins like ricin and saporin function by inactivating ribosomes,[9,10] and the bacterial toxins *Pseudomonas* exotoxin (PE) and diphtheria toxin (DT) inhibit protein synthesis by ADP ribosylation of elongation factor 2.[11,12] The bacterial toxins are ideal for fusing with ligands to make recombinant toxins, because they are produced by bacteria as single chains, containing a binding domain that may be exchanged for a ligand for tumor cells.[13] The general public in developed countries is routinely immunized against DT, but not PE. Nevertheless, both toxins have been employed successfully to treat hematologic tumors in patients, particularly when the patients are too immunosuppressed to be affected by the immunogenicity of the toxin. The use of DT in hematologic and solid tumors is discussed elsewhere in this volume. The remainder of this review focuses on the use of PE in treating hematologic malignancies.

Structure and function of PE-based toxins

Detailed models for mechanisms of action of PE are reviewed elsewhere in this volume. Briefly, as shown in Figure 1.1, PE has a binding domain (domain Ia) at the amino terminus, an enzymatic domain (domain III) at the carboxyl terminus, and a domain in the middle (domain II), which induces translocation of the enzymatic domain to the cytoplasm of the target cell.[14,15] The toxin is proteolytically cleaved by furin within the translocating domain between amino acids Arg279 and Gly280 so that the enzymatic domain becomes separate from the binding domain. However, the two parts are held together by a disulfide bond until it is reduced inside the cell.[16] The enzymatic domain then translocates to the cytosol without the binding domain, where it ADP-ribosylates elongation factor 2. This kills the cell by inhibition of protein synthesis or by induction of apoptosis.[11,17] In recombinant toxins the ligand replaces the binding domain at the amino terminus of PE.

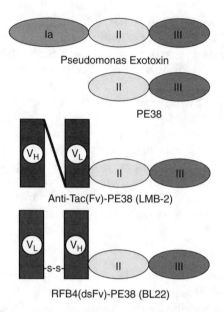

Figure 1.1 **Schematic structure of *Pseudomonas* exotoxin (PE) and recombinant toxins**. PE is a single-chain protein 613 amino acids long and contains 3 functional domains.[14,15] Domain Ia (amino acids 1–252) is the binding domain, domain II (amino acids 253–364) is the translocating domain, and domain III (amino acids 400–613) contains the ADP ribosylating enzyme, which inactivates elongation factor 2 (EF-2) in the cytosol and results in cell death.[99] Domain Ib (not shown) separates domains II and III and contains amino acids 365–399. PE38 is a 38 kDa truncated form of PE devoid of both domain Ia and amino acids 365–380 of domain Ib. The single-chain recombinant immunotoxin anti-Tac(Fv)-PE38 (LMB-2) contains the variable heavy domain (V_H) of the anti-Tac MAb fused via the peptide linker $(G_4S)_3$ to the variable light domain (V_L), which in turn is fused to PE38. The recombinant immunotoxin RFB4(dsFv)-PE38 (BL22) is composed of the V_L from the MAb RFB4 disulfide bonded to a fusion of V_H with PE38. The disulfide bond connecting V_H and V_L is formed between two cysteine residues replacing Arg44 of V_H and Gly100 of V_L

Fragments of PE used for connecting to ligands

Commonly used PE fragments without binding domains include amino acids 254–613, called PE40,[14] and 253–364 and 381–613 of PE, called PE38.[18,19] The amino acids 365–380, which are absent in PE38, contain a disulfide bond that is not necessary for cytotoxic activity. PE38KDEL is a mutant of PE38 with increased activity in which the last five amino acids REDLK (arginine-glutamate-aspartate-leucine-lysine) are mutated to KDEL (lysine-aspartate-glutamate-leucine).[19–21] The proteolytic cleavage of single-chain toxin within a disulfide loop is a special quality of PE and DT, which allows either of these toxins to be used to make single-chain recombinant toxins.[13]

Without the proteolytic cleavage site, the ligand would not be able to separate from the enzymatic domain and translocation to the cytosol would be severely impaired. Moreover, without the disulfide loop, the binding domain might separate from the enzymatic domain outside the cell and prevent selective targeting. On the other hand, if the ligand is to be conjugated to PE through a disulfide bond, the proteolytic cleavage site of PE is not needed. Instead, the ligand can be conjugated to a 35 kDa fragment of PE, which is identical to the translocated fragment of PE38 (amino acids 280–364 and 381–613) except that methionine rather than Glycine is present at the amino terminus.[22] This fragment is termed PE35, and contains a single cysteine residue at position 287.

PE-based toxins targeting the interleukin-2 receptor

Interleukin-2 receptor as a target for immunotoxins

The interleukin-2 receptor (IL2R) is present on lymphoma, leukemia, Hodgkin's disease, and also the activated T-cells mediating autoimmune disorders.[23-30] The IL2R is made up of the α subunit (p55, CD25, or Tac), which binds IL2 with low affinity (Kd ~ 10^{-8} M), and the complex of β and γ subunits, which bind IL2 with intermediate affinity (Kd ~ 10^{-9} M). All three subunits together bind IL2 with high affinity (Kd ~ 10^{-11} M).[31]

Targeting IL2R with IL2-PE toxins

Originally PE40 was targeted to IL2R-expressing cells using human IL2.[32] IL2-PE40 was cytotoxic to activated murine T-cells that mediate autoimmune disease, and was active in a variety of animal models of autoimmune disease.[32-38] IL2-PE40 was much less cytotoxic for human activated T-cells. A more active form of IL-2 toxin was IL2-PE66[4Glu], wherein a full-length form of the toxin was used that contained mutations in domain Ia that prevented binding to normal cells. Compared to IL2-PE40, IL2-PE66[4Glu] was more cytotoxic toward activated human T-cells and was also more toxic to mice.[39]

Early immunotoxins targeting CD25

It was recognized that many tumor types, including adult T-cell leukemia (ATL) and chronic lymphocytic leukemia (CLL), express much more CD25 than other IL2R subunits.[25,27] Moreover, in contrast to the low affinity binding of IL2 to CD25 alone, the anti-CD25 MAb anti-Tac binds CD25 with high affinity (Kd ~ 10^{-10} M). Despite earlier evidence from anti-Tac internalization studies,[41] CD25 alone internalizes bound ligand.[19,42] Originally PE was targeted to CD25 in its full-length form via a chemical conjugate of murine anti-Tac and whole PE. The PE portion was treated with iminothiolane, which modifies lysine residues in domain Ia and thereby decreases the binding of PE to its receptor[43] and facilitated conjugation of the toxin to anti-Tac. Anti-Tac-PE inhibited protein synthesis in an HTLV-I positive T-cell line HUT-102 by 50% (IC_{50}) at 1.2 ng/ml (5 pM), compared to 90–880 ng/ml for CD25 negative cells.[44] Anti-Tac-PE was selective enough to eliminate activated T-cells *in vitro*.[45] In a pilot trial, anti-Tac-PE was tested in several patients with adult T-cell leukemia. This first-generation immunotoxin was not clinically active, probably in part due to increased toxicity from binding to normal cells via the PE binding domain, and in part due to immunogenicity from both the toxin and MAb. The chemical conjugate anti-Tac-Lys-PE40, which does not contain the PE binding domain, was less cytotoxic than anti-Tac-PE toward HUT-102 cells, with an IC_{50} of 13 pM compared to 5 pM; but the nonspecific cytotoxicity toward CD25-negative cells improved 6 to 100-fold.[44,46] PE35 was chemically conjugated through its unique cysteine residue at position 287 to anti-Tac.[22] Anti-Tac-PE35 is unique among other anti-CD25 immunotoxins in that proteolytic processing of PE is not needed for cytotoxicity. However, anti-Tac-PE35 was found to have an IC_{50} on HUT-102 cells of ~20 pM, which was not improved over that of the chemical conjugate anti-Tac-PE40.[22,44,46]

Recombinant immunotoxins targeting CD25

To target CD25+ cells with a recombinant protein, a single-chain Fv was produced from the variable domains of anti-Tac (Figure 1.1) and this was fused to truncated forms of PE. A single-chain Fv molecule composed of one variable domain connected to the next through a peptide linker is only 26 kDa in size, less than 20% of the size of an IgG. Anti-Tac(Fv) was constructed by fusing V_H to V_L via the 15 amino acid linker $(G_4S)_3$ and the resulting Fv fragment of anti-Tac was fused to PE40.[47] Anti-Tac(Fv)-PE40 retained 1/3 of the binding affinity of anti-Tac IgG. This recombinant immunotoxin was extremely cytotoxic with an IC_{50} of 0.15 ng/ml (~2 pM) toward HUT-102 cells[47] and 0.05–0.1 ng/ml toward activated human T-cells.[19, 48] Anti-Tac(Fv)-PE40 and its slightly smaller version anti-Tac(Fv)-PE38 were very cytotoxic toward fresh malignant cells from 46 patients with ATL.[19,49–51] Anti-Tac(Fv)-PE38KDEL is a mutant of Anti-Tac(Fv)-PE38, which contains the KDEL sequence at the carboxyl terminus instead of the native REDLK sequence of PE. Anti-Tac(Fv)-PE38 and anti-Tac(Fv)-PE38KDEL were also very cytotoxic toward CD25+ HCL cells, and were also cytotoxic (albeit with higher IC50s) toward CLL cells.[52] Anti-Tac(Fv)-PE38 showed antitumor activity against CD25-expressing human tumors in mice.[53,54] Anti-Tac(Fv)-PE38KDEL was very cytotoxic toward fresh cells from patients with CLL and PLL, although its liver toxicity in mice is also increased compared to anti-Tac(Fv)-PE38.[28,52,53,55] A recombinant immunotoxin was made containing PE40 and anti-Tac(Fab), which due to its increased size had a markedly increased half-life in mice.[56] When preclinical activity, production yields, and molecular size were considered for these recombinant immunotoxins, it was decided to continue development of anti-Tac(Fv)-PE38. This 63 kDa single-chain recombinant immunotoxin was called LMB-2 (Figure 1.1).

Toxicology and biodistribution of LMB-2 in animals

For toxicology evaluation, primate studies were important, since primate but not murine CD25 binds anti-Tac. In GLP studies, anti-Tac(Fv)-PE38 was administered to four Cynomolgus monkeys at 20 mg i.v. q.o.d. × 3, and this was determined to be a no-effect dose. Monkeys tolerated anti-Tac(Fv)-PE38 at doses of 300, 750, and 1000 mg/kg i.v.q.o.d. × 3 with only loss of appetite and reversible transaminase elevations. Pharmacokinetic studies with anti-Tac(Fv)-PE38 at 750 mg/kg i.v. showed a biphasic disappearance from serum with a $T_{1/2\alpha}$ of 58 min and a $T_{1/2\beta}$ of 170 min, and an area under the curve (AUC) of 2700 µg-min/ml. To determine the biodistribution of LMB-2 into human tumors and normal organs, mice bearing CD25+ human tumors were injected with {[125]I}-LMB-2. It was determined that CD25+ tumors selectively concentrated LMB-2 at 90 and 360 min time points, and that uptake in kidney exceeded that in liver in terms of percentage of injected dose per gram of tissue.[57,58] It was also determined that less than 1000 molecules of LMB-2 needed to bind *in vivo* to induce complete regressions. When such experiments were repeated with {[111]In}-LMB-2, which is more accurate for measuring LMB-2 uptake, it was found that LMB-2 uptake into CD25 tumors was significant even at 96 h.[59]

Clinical activity of LMB-2 in patients with CD25+ malignancies

LMB-2 was administered to 35 patients with chemotherapy-resistant leukemia, lymphoma, and Hodgkin's disease (HD). The dose levels were 2, 6, 10, 20, 30, 40, 50, and 63 µg/kg every other day for three doses (q.o.d. ×3). Eight major responses were observed, including one complete remission (CR) and seven partial responses (PR); all major responses were observed in the twenty patients receiving > 60 µg/kg/cycle for a response rate of 40% in this high dose group. Out of four patients with HCL treated, 100% responded, including one CR.[60] The patient with CR had resolution of severe pancytopenia and eradication of circulating malignant cells. The hemoglobin was as low as 3.8 prior to treatment (prior to transfusions), and has remained normal nearly 5 years after LMB-2. PRs were achieved in patients with CLL, ATL, CTCL, and HD.[61] Dose-limiting toxicity (DLT) occurred at the highest dose level (63 µg/kg q.o.d. ×3), where one patient had a reversible grade IV cardiomyopathy and another patient had reversible grade IV transaminase elevations. The maximum tolerated dose (MTD) was 40 µg/kg q.o.d. ×3. Low grade toxicity at lower dose levels included fever and transaminase elevations. Transaminase elevations in mice appear to be mediated at least in part by cytokines.[62,63] Immunogenicity from LMB-2 was lower than expected, with only 6 of 35 patients prevented from further treatment because of neutralizing antibodies after cycle one.[61] CLL patients never made neutralizing antibodies even after a total of 16 cycles administered to 8 patients. Three of these patients did develop low levels of nonneutralizing anti-PE38 antibodies without HAMA. Neutralizing antibody formation was higher in HD; out of 11 patients with HD, 3 patients made high levels of neutralizing antibodies after 1 cycle and 2 patients made high levels after two or three cycles. The median half-life of LMB-2 at the MTD was about 4 h, and its disappearance from the plasma was usually monoexponential, particularly in the absence of circulating malignant cells.

Future development of anti-CD25 recombinant immunotoxins

A phase II trial of LMB-2 is planned in patients with CD25+ hematologic malignancies, such as CLL and CTCL. Phase I trials are planned in pediatric CD25+ malignancies and also for the prevention or treatment of GVHD in patients undergoing high-risk allotransplantation.[64,65] To improve stability of the single-chain Fv ligand, which is flexible and prone to aggregation, cysteine residues were engineered into the framework region of anti-Tac(Fv)-PE38KDEL and the linker between V_H and V_L was replaced by a more stable disulfide bond. The disulfide-stabilized recombinant immunotoxin anti-Tac(dsFv)-PE38KDEL was no longer a single-chain protein, since it contained V_H as a separate protein disulfide-bonded to a fusion of V_L with PE38KDEL. Nevertheless, it could be made without chemical conjugation because both subunits, as reduced and denatured bacterial inclusion body protein, renature to the active immunotoxin when added to the refolding buffer. The disulfide stabilization did not impair binding, cytotoxic activity, or antitumor activity but it did improve stability at 37°C.[66,67] It was found that the isoelectric point (pI) of the Fv of LMB-2 could be lowered by mutagenizing residues in the framework regions, and that this led to a decrease in animal toxicity and improvement in the therapeutic index.[62,63,68] Further preclinical development of this low-pI mutant of LMB-2 is needed prior to clinical testing of this new agent. To address issues of immunogenicity and half-life, LMB-2 was PEGylated and found to retain cytotoxicity and display enhanced antitumor activity.[69] Like the MAb Anti-Tac, which was converted to LMB-2, the anti-CD25 MAb RFT5 was converted to the recombinant immunotoxin RFT5(scFv)-ETA'. The truncated toxin ETA' is almost identical to PE40 and contains amino acids 252–613 of PE. RFT5(scFv)-ETA' induced antitumor activity in SCID mice bearing disseminated human HD.[70,71]

PE-based toxins targeting CD22

CD22 is a cell-adhesion molecule expressed by mature and malignant B-cells, which is involved in antibody responses and cell-lifespan.[72,73] Truncated PE was first targeted to CD22-bearing cells using chemical conjugation to the LL2 antibody or Fab' fragment, and the immunotoxins showed antitumor activity against solid Burkitt's lymphoma xenografts in mice.[22,74] Murine LL2 when converted to a single-chain Fv was not stable enough to result in a recombinant immunotoxin with acceptable properties. A different CD22 MAb, RFB4, had previously been shown to effectively target ricin derivatives to CD22+ cells, and chemical conjugates between ricin and RFB4 or RFB4-Fab' showed major responses in up to 30% of patients with B-cell malignancies.[75–79] RFB4 was chemically conjugated to truncated forms of PE[80] and the RFB4-immunotoxins were cytotoxic toward CD22+ cells lines.

Targeting CD22 with recombinant immunotoxin

To construct a recombinant immunotoxin targeting CD22, the variable domains of RFB4 were connected via a peptide linker and then connected to PE38.[81] To improve the stability of RFB4(Fv)-PE38, the variable domains were connected by a disulfide bond instead of a peptide linker, and VH was fused to PE38, resulting in RFB4(dsFv)-PE38 (Figure 1.1).[82] The disulfide bond is between cysteine residues replacing framework residues Arg44 of VH and Gly100 of V_L. This protein was called BL22 and was developed further for the treatment of CD22+ malignancies.

Preclinical development of BL22

To determine the antitumor activity and animal toxicology of BL22, the recombinant disulfide-stabilized immunotoxin was administered to nude mice bearing human CD22+ tumors and to monkeys. It was found that complete regressions in mice of human CD22+ B-cell lymphoma xenografts were observed at plasma levels that could be tolerated in Cynomolgus monkeys.[83] Fresh leukemic cells obtained from patients with CLL and NHL were also tested in tissue culture with BL22 and found to be very sensitive.[84] This study, which showed specific killing of such cells, was important for preclinical development because malignant cells freshly obtained from patients typically display far fewer CD22 sites/cells compared to cell lines.

Phase I testing of BL22 in patients with B-cell malignancies

BL22 has been tested in a phase I trial of hematologic malignancies. Results on the first 31 patients have been reported, particularly in 16 of these patients with HCL.[85] In addition to the 16 HCL patients, 4 of the 31 patients had NHL and 11 had CLL. Like LMB-2, patients were dosed by 30 min infusion q.o.d. ×3. Patients without progressive disease or neutralizing antibodies to the toxin could be retreated at 3-week intervals. Of the 16 HCL patients, 13 had classic HCL and 3 had HCLv, a poorer prognosis form of HCL in which the response to purine analog therapy is very poor. All HCL patients were pretreated with 1 to 6 separate courses of cladribine, which is considered the most effected treatment since it can induce CR in up to 85% of patients after a single 1-week course.[86–89] Patients often had one or several courses of prior interferon, pentostatin, and rituximab and about half had prior splenectomy. Dose levels included 3, 6, 10, 20, 30, 40, and 50 ug/kg q.o.d. ×3, and patients often were retreated with a variety of different dose levels.

Activity of BL22 in patients with cladribine-resistant HCL

A total of 87 cycles of BL22 were administered to 16 patients, with up to 15 cycles/patient. The response rate in HCL was 81%, with 11 CRs and 2 PRs (12%) out of 16 evaluable patients.[85] Two patients had marginal responses with 98 to 99.5% reductions in circulating HCL counts but with persistent lymph node masses. All 3 of the HCLv patients achieved CR. Out of the 11 CRs, 6 achieved CR after cycle 1, and 5 had CR after cycles 2 to 9. Only 1 out of 11 CRs had minimal residual disease (MRD) in the bone marrow biopsy by immunohistochemistry, which is know to be a risk factor for early relapse.[90] A different patient had MRD in the blood by flow cytometry. The other 9 CRs had no MR in the bone marrow biopsy or in the blood. Baseline cytopenias including thrombocytopenia and granulocytopenia resolved after the first cycle. Within the follow-up time of 4 to 22 (median 12) months, relapse was documented in 3 patients after 7 to 12 months, and all 3 patients reachieved CR with additional BL22. High levels of neutralizing antibodies were observed in 3 patients after cycles 1 to 5. Plasma levels in patients with high disease burden increased as they responded, indicating that the hairy cells constituted a sink for the BL22 because of their high CD22 expression. In the HCL patients, serious toxicity included a cytokine release syndrome in one patient with fever, hypotension, bone pain, and weight gain (VLS) without pulmonary edema, which resolved within 3 days. Also, 2 patients had a completely reversible hemolytic uremic syndrome (HUS), confirmed by renal biopsy. HUS presented clinically with hematuria and hemoglobinuria by day 8 of cycle 2 in each case. These patients required 6 to 10 days of plasmapheresis but not dialysis for complete resolution of renal function and correction of thrombocytopenia and anemia. Both patients achieved CR and in both cases all preexisting cytopenias resolved as well as those related to HUS. These interim results thus indicated that BL22 is the first agent since purine analogs that can induce CR in the majority of patients with HCL, and may be the only agent that can induce CR in the majority of patients with chemotherapy resistant HCL, particularly HCLv. The cause of HUS in patients receiving BL22 is not known, but since HUS has not been observed in over 100 patients receiving PE38 fused to ligands other than RFB4(dsFv), the mechanism may in part be CD22-related.

Further clinical development of BL22

A phase II trial of BL22 is planned in HCL, and a phase I trial in CLL/NHL is planned to more fully explore the activity of BL22 in those diseases. In addition, *ex vivo* studies have indicated that pediatric acute lymphoblastic leukemia cells are usually CD22+ and sensitive to BL22. A phase I trial of BL22 in pediatric B-cell leukemia is planned.

Improved form of BL22

To improve the targeting of PE to CD22+ cells, the variable domains of BL22 were mutated to improve affinity. The method used was based on the observation that somatic mutations in antibodies frequently occur in certain sequences that appear in each antibody gene.[91,92] These so called "hot spots" were randomized within the CDR3 domain of the RFB4 V_H, and the resulting small libraries of single-chain Fvs screened by phage display.[93] It was determined that a mutant containing the sequence THW instead of SSY at positions 100, 100a, and 100b was much more cytotoxic than wild type immunotoxin. A new high-affinity mutant of BL22, termed HA22, is currently undergoing preclinical development for treating CD22+ malignancies, particularly CLL, where the low CD22 expression may prevent optimal response to BL22.

PE-based toxins targeting CD30

Hodgkin's disease Reed Sternberg (HRS) cells consistently express high numbers of CD30 antigens, a member of the tumor necrosis factor (TNF)/nerve growth factor (NGF) receptor superfamily.[94] Tumors from patients with non-Hodgkin's lymphoma (NHL) are often CD30+, particularly in anaplastic large cell lymphomas (ALCL), which make up 5% of all NHL.[95] Mediastinal large B-cell lymphomas are CD30+ in 69% of cases.[96] Because of its restricted expression in only a small subset of normal lymphocytes, the CD30 antigen is a potential candidate for selective targeting by CD30-specific antibodies. Recombinant PE-based immunotoxins targeting CD30 were produced by immunizing mice with DNA encoding human CD30, using the spleens to construct an Fv phage display library, and isolating Fvs that bound to soluble CD30.[97] One of the anti-CD30(Fv)-PE38KDEL molecules produced by this approach showed significant cytotoxicity toward CD30+ cell lines and displayed antitumor activity against CD30+ tumors in mice. In a second approach, spleens from CD30 DNA-immunized mice were used to produce a panel of hybridomas, which were each produced and screened for activity to CD30. A total of four anti-CD30 disulfide-stabilized recombinant immunotoxins were produced, two of which had good cytotoxicity to CD30+ cell lines.[98] These latter recombinant immunotoxins were more cytotoxic than those produced by phage display.

Conclusions

Recombinant immunotoxins have proven useful in hematologic tumors. Although they have not been tested head-to-head with their conventional immunotoxin counterparts in humans or even in mice, their advantages, such as ease of production, small size, <50% immunogenicity in leukemias and lymphomas, and short half-life without life-threatening VLS, are notable. Recombinant immunotoxins may be further improved by protein engineering to increase cytotoxicity and decrease undesirable toxicity. They constitute a potentially useful treatment modality alongside other forms of surface-targeted therapy, since: (1) they do not require immune mechanisms for cell death; (2) they can induce cell death without apoptosis even though apoptosis does facilitate cytotoxicity;[17] (3) they appear not to cause cumulative toxicity like radioimmunotherapy; and finally, (4) resistance to toxin appears much more rare than resistance to chemotherapy. Major obstacles to success remain, such as (1) immunogenicity in a significant proportion of patients, (2) serious toxicities such as hemolytic uremic syndrome and cytokine release syndrome, which are specific for the ligand used, and (3) limited plasma lifetime, which may reduce response rates, particularly in the presence of bulky disease. Approaches under consideration or development to address these issues include (1) administration by continuous infusion to increase plasma lifetime and tumor penetration, (2) engineering new improved ligands with more specificity and less toxicity, and (3) combination with chemotherapy and other agents to increase efficacy and decrease immunogenicity. It is anticipated that significant progress will continue for the development of PE-based recombinant immunotoxins for the treatment of hematologic malignancies.

References

1. Jemal, A., Murray, T., Samuels, A., Ghafoor, A., Ward, E., and Thun, M.J. Cancer statistics, 2003. *CA Cancer J. Clin.*, *53*: 5–26, 2003.
2. Akhtar, S. and Maghfoor, I. Rituximab plus CHOP for diffuse large-B-cell lymphoma. *N. Engl. J. Med.*, *346*: 1830–1831; discussion 1830–1831, 2002.
3. Keating, M.J., Flinn, I., Jain, V., Binet, J.L., Hillmen, P., Byrd, J., Albitar, M., Brettman, L., Santabarbara, P., Wacker, B., and Rai, K.R. Therapeutic role of alemtuzumab (Campath-1H) in patients who have failed fludarabine: results of a large international study. *Blood*, *99*: 3554–3561, 2002.
4. Cheson, B. Bexxar (Corixa/GlaxoSmithKline). *Curr. Opin. Investig. Drugs*, *3*: 165–170, 2002.
5. Alcindor, T. and Witzig, T.E. Radioimmunotherapy with Yttrium-90 Ibritumomab Tiuxetan for Patients with relapsed CD20+ B-cell non-Hodgkin's lymphoma. *Curr. Treat. Options. Oncol.*, *3*: 275–282, 2002.
6. Sievers, E.L., Appelbaum, F.R., Spielberger, R.T., Forman, S.J., Flowers, D., Smith, F.O., Shannon-Dorcy, K., Berger, M.S., and Bernstein, I.D. Selective ablation of acute myeloid leukemia using antibody-targeted chemotherapy: a phase I study of an anti-CD33 calicheamicin immunoconjugate. *Blood*, *93*: 3678–3684, 1999.
7. Nabhan, C. and Tallman, M.S. Early phase I/II trials with Gemtuzumab Ozogamicin (Mylotarg(R)) in acute myeloid leukemia. *Clin. Lymphoma*, *2 Suppl. 1*: S19–23, 2002.
8. Naito, K., Takeshita, A., Shigeno, K., Nakamura, S., Fujisawa, S., Shinjo, K., Yoshida, H., Ohnishi, K., Mori, M., Terakawa, S., and Ohno, R. Calicheamicin-conjugated humanized anti-CD33 monoclonal antibody (Gemtuzumab zogamicin, CMA-676) shows cytocidal effect on CD33-positive leukemia cell lines, but is inactive on P-glycoprotein-expressing sublines. *Leukemia*, *14*: 1436–1443, 2000.

9. Endo, Y., Mitsui, K., Motizuki, M., and Tsurugi, K. The mechanism of action of ricin and related toxic lectins on eukaryotic ribosomes. *J. Biol. Chem., 262*: 5908–5912, 1987.

10. Zamboni, M., Brigotti, M., Rambelli, F., Montanaro, L., and Sperti, S. High pressure liquid chromatographic and fluorimetric methods for the determination of adenine released from ribosomes by ricin and gelonin. *Biochem. J., 259*: 639–643, 1989.

11. Carroll, S.F. and Collier, R.J. Active site of *Pseudomonas aeruginosa* exotoxin A. Glutamic acid 553 is photolabeled by NAD and shows functional homology with glutamic acid 148 of diphtheria toxin. *J. Biol. Chem., 262*: 8707–8711, 1987.

12. Bell, C.E. and Eisenberg, D. Crystal structure of diphtheria toxin bound to nicotinamide adenine dinucleotide. *Biochemistry, 35*: 1137–1149, 1996.

13. Kreitman, R.J. Getting plant toxins to fuse. *Leukemia Res., 21*: 997–999, 1997.

14. Hwang, J., FitzGerald, D.J., Adhya, S., and Pastan, I. Functional domains of Pseudomonas exotoxin identified by deletion analysis of the gene expressed in *E. coli. Cell, 48*: 129–136, 1987.

15. Allured, V.S., Collier, R.J., Carroll, S.F., and McKay, D.B. Structure of exotoxin A of *Pseudomonas aeruginosa* at 3.0 Angstrom resolution. *Proc. Natl. Acad. Sci. USA, 83*: 1320–1324, 1986.

16. McKee, M.L. and FitzGerald, D.J. Reduction of furin-nicked Pseudomonas exotoxin A: an unfolding story. *Biochemistry, 38*: 16507–16513, 1999.

17. Keppler-Hafkemeyer, A., Kreitman, R.J., and Pastan, I. Apoptosis induced by immunotoxins used in the treatment of hematologic malignancies. *Int. J. Cancer, 87*: 86–94, 2000.

18. Siegall, C.B., Chaudhary, V.K., FitzGerald, D.J., and Pastan, I. Functional analysis of domains II, Ib, and III of *Pseudomonas* exotoxin. *J. Biol. Chem., 264*: 14256–14261, 1989.

19. Kreitman, R.J., Batra, J. K., Seetharam, S., Chaudhary, V.K., FitzGerald, D.J., and Pastan, I. Single-chain immunotoxin fusions between anti-Tac and *Pseudomonas* exotoxin: relative importance of the two toxin disulfide bonds. *Bioconjugate Chem., 4*: 112–120, 1993.

20. Kreitman, R.J. and Pastan, I. Importance of the glutamate residue of KDEL in increasing the cytotoxicity of *Pseudomonas* exotoxin derivatives and for increased binding to the KDEL receptor. *Biochem. J., 307*: 29–37, 1995.

21. Seetharam, S., Chaudhary, V.K., FitzGerald, D., and Pastan, I. Increased cytotoxic activity of *Pseudomonas* exotoxin and two chimeric toxins ending in KDEL. *J. Biol. Chem., 266*: 17376–17381, 1991.

22. Theuer, C.P., Kreitman, R.J., FitzGerald, D.J., and Pastan, I. Immunotoxins made with a recombinant form of *Pseudomonas* exotoxin A that do not require proteolysis for activity. *Cancer Res., 53*: 340–347, 1993.

23. Nasu, K., Said, J., Vonderheid, E., Olerud, J., Sako, D., and Kadin, M. Immunopathology of cutaneous T-cell lymphomas. *Am. J. Pathol., 119*: 436–447, 1985.

24. Foss, F.M., Borkowski, T.A., Gilliom, M., Stetler-Stevenson, M., Jaffe, E.S., Figg, W.D., Tompkins, A., Bastian, A., Nylen, P., Woodworth, T., Udey, M.C., and Sausville, E.A. Chimeric fusion protein toxin $DAB_{486}IL-2$ in advanced mycosis fungoides and the sezary syndrome: correlation of activity and interleukin-2 receptor expression in a phase II study. *Blood, 84*: 1765–1774, 1994.

25. Kodaka, T., Uchiyama, T., Ishikawa, T., Kamio, M., Onishi, R., Itoh, K., Hori, T., Uchino, H., Tsudo, M., and Araki, K. Interleukin-2 receptor β-chain (p70–75)

expressed on leukemic cells from adult T cell leukemia patients. *Jpn. J. Cancer Res.*, *81*: 902–908, 1990.

26. Robbins, B.A., Ellison, D.J., Spinosa, J.C., Carey, C.A., Lukes, R.J., Poppema, S., Saven, A., and Piro, L. D. Diagnostic application of two-color flow cytometry in 161 cases of hairy cell leukemia. *Blood*, *82*: 1277–1287, 1993.

27. Yagura, H., Tamaki, T., Furitsu, T., Tomiyama, Y., Nishiura, T., Tominaga, N., Katagiri, S., Yonezawa, T., and Tarui, S. Demonstration of high-affinity inter-leukin-2 receptors on B-chronic lymphocytic leukemia cells: functional and structural characterization. *Blut*, *60*: 181–186, 1990.

28. Kreitman, R.J. and Pastan, I. Recombinant single-chain immunotoxins against T and B cell leukemias. *Leuk. Lymphoma*, *13*: 1–10, 1994.

29. Agnarsson, B.A. and Kadin, M.E. The immunophenotype of Reed-Sternberg cells. A study of 50 cases of Hodgkin's disease using fixed frozen tissues. *Cancer*, *63*: 2083–2087, 1989.

30. Sheibani, K., Winberg, C. D., Velde, S.V.D., Blayney, D.W., and Rappaport, H. Distribution of lymphocytes with interleukin-2 receptors (TAC antigens) in reactive lymphoproliferative processes, Hodgkin's disease, and non-Hodgkin's lymphomas: an immunohistologic study of 300 cases. *Am. J. Pathol.*, *127*: 27–37, 1987.

31. Taniguchi, T. and Minami, Y. The IL2/IL-2 receptor system: a current overview. *Cell*, *73*: 5–8, 1993.

32. Lorberboum-Galski, H., FitzGerald, D., Chaudhary, V., Adhya, S., and Pastan, I. Cytotoxic activity of an interleukin 2-*Pseudomonas* exotoxin chimeric protein produced in *Escherichia coli*. *Proc. Natl. Acad. Sci. USA*, *85*: 1922–1926, 1988.

33. Lorberboum-Galski, H., Barrett, L.V., Kirkman, R.L., Ogata, M., Willingham, M.C., FitzGerald, D.J., and Pastan, I. Cardiac allograft survival in mice treated with IL-2-PE40. *Proc. Natl. Acad. Sci. USA*, *86*: 1008–1012, 1989.

34. Lorberboum-Galski, H., Kozak, R.W., Waldmann, T.A., Bailon, P., FitzGerald, D.J.P., and Pastan, I. Interleukin 2 (IL2) PE40 is cytotoxic to cells displaying either the p55 or p70 subunit of the IL2 receptor. *J. Biol. Chem.*, *263*: 18650–18656, 1988.

35. Lorberboum-Galski, H., Lafyatis, R., Case, J.P., FitzGerald, D., Wilder, R.L., and Pastan, I. Administration of IL-2-PE40 via osmotic pumps prevents adjuvant induced arthritis in rats. Improved therapeutic index of IL-2-PE40 administered by continuous infusion. *Intl. J. Immunopharmac.*, *13*: 305–315, 1991.

36. Case, J.P., Lorberboum-Galski, H., Lafyatis, R., FitzGerald, D., Wilder, R.L., and Pastan, I. Chimeric cytotoxin IL2-PE40 delays and mitigates adjuvant-induced arthritis in rats. *Proc. Natl. Acad. Sci. USA*, *86*: 287–291, 1989.

37. Herbort, C.P., Smet, M. D.d., Roberge, F.G., Nussenblatt, R.B., FitzGerald, D., Lorberboum-Galski, H., and Pastan, I. Treatment of corneal allograft rejection with the cytotoxin IL-2-PE40. *Transplantation*, *52*: 470–474, 1991.

38. Rose, J.W., Lorberboum-Galski, H., FitzGerald, D., McCarron, R., Hill, K.E., Townsend, J.J., and Pastan, I. Chimeric cytotoxin IL2-PE40 inhibits relapsing experimental allergic encephalomyelitis. *J. Neuroimmunol.*, *32*: 209–217, 1991.

39. Lorberboum-Galski, H., Garsia, R.J., Gately, M., Brown, P.S., Clark, R.E., Waldmann, T.A., Chaudhary, V.K., FitzGerald, D.J.P., and Pastan, I. IL2-PE66[4Glu], a new chimeric protein cytotoxic to human activated T lymphocytes. *J. Biol. Chem.*, *265*: 16311–16317, 1990.

40. Uchiyama, T.A., Broder, S., and Waldmann, T.A. A monoclonal antibody (anti-Tac) reactive with activated and functionally mature human T cells.

I. Production of anti-Tac monoclonal antibody and distribution of Tac (+) cells. *J. Immunol., 126*: 1393–1397, 1981.

41. Weissman, A.M., Harford, J.B., Svetlik, P.B., Leonard, W.L., Depper, J.M., Waldmann, T.A., Greene, W.C., and Klausner, R.D. Only high-affinity receptors for interleukin 2 mediate internalization of ligand. *Proc. Natl. Acad. Sci. USA, 83*: 1463–1466, 1986.

42. Chaudhary, V.K., Gallo, M.G., FitzGerald, D.J., and Pastan, I. A recombinant single-chain immunotoxin composed of anti-Tac variable regions and a truncated diphtheria toxin. *Proc. Natl. Acad. Sci. USA, 87*: 9491–9494, 1990.

43. FitzGerald, D.J.P., Waldmann, T.A., Willingham, M.C., and Pastan, I. Pseudomonas exotoxin-Anti-Tac: cell specific immunotoxin active against cells expressing the human T cell growth factor receptor. *J. Clin. Invest., 74*: 966–971, 1984.

44. Kondo, T., FitzGerald, D., Chaudhary, V.K., Adhya, S., and Pastan, I. Activity of immunotoxins constructed with modified *Pseudomonas* exotoxin A lacking the cell recognition domain. *J. Biol. Chem., 263*: 9470–9475, 1988.

45. Kozak, R.W., Fitzgerald, D.P., Atcher, R.W., Goldman, C.K., Nelson, D.L., Gansow, O.A., Pastan, I., and Waldmann, T.A. Selective elimination *in vitro* of alloresponsive T cells to human transplantation antigens by toxin or radionuclide conjugated anti-IL-2 receptor (Tac) monoclonal antibody. *J. Immunol., 144*: 3417–3423, 1990.

46. Batra, J.K., Jinno, Y., Chaudhary, V.K., Kondo, T., Willingham, M.C., FitzGerald, D.J., and Pastan, I. Antitumor activity in mice of an immunotoxin made with anti-transferrin receptor and a recombinant form of *Pseudomonas* exotoxin. *Proc. Natl. Acad. Sci. USA, 86*: 8545–8549, 1989.

47. Chaudhary, V.K., Queen, C., Junghans, R.P., Waldmann, T.A., FitzGerald, D.J., and Pastan, I. A recombinant immunotoxin consisting of two antibody variable domains fused to *Pseudomonas* exotoxin. *Nature, 339*: 394–397, 1989.

48. Batra, J.K., FitzGerald, D., Gately, M., Chaudhary, V.K., and Pastan, I. Anti-Tac(Fv)-PE40: a single chain antibody *Pseudomonas* fusion protein directed at interleukin 2 receptor bearing cells. *J. Biol. Chem., 265*: 15198–15202, 1990.

49. Kreitman, R.J., Chaudhary, V.K., Waldmann, T., Willingham, M.C., FitzGerald, D.J., and Pastan, I. The recombinant immunotoxin anti-Tac(Fv)-*Pseuodomonas* exotoxin 40 is cytotoxic toward peripheral blood malignant cells from patients with adult T-cell leukemia. *Proc. Natl. Acad. Sci. USA, 87*: 8291–8295, 1990.

50. Kreitman, R.J., Chaudhary, V.K., Waldmann, T.A., Hanchard, B., Cranston, B., FitzGerald, D.J.P., and Pastan, I. Cytotoxic activities of recombinant immunotoxins composed of *Pseudomonas* toxin or diphtheria toxin toward lymphocytes from patients with adult T-cell leukemia. *Leukemia, 7*: 553–562, 1993.

51. Saito, T., Kreitman, R.J., Hanada, S.-i., Makino, T., Utsunomiya, A., Sumizawa, T., Arima, T., Chang, C.N., Hudson, D., Pastan, I., and Akiyama, S.-i. Cytotoxicity of recombinant Fab and Fv immunotoxins on adult T-cell leukemia lymph node and blood cells in the presence of soluble interleukin-2 receptor. *Cancer Res., 54*: 1059–1064, 1994.

52. Robbins, D.H., Margulies, I., Stetler-Stevenson, M., and Kreitman, R.J. Hairy cell leukemia, a B-cell neoplasm which is particularly sensitive to the cytotoxic effect of anti-Tac(Fv)-PE38 (LMB-2). *Clin. Cancer Res., 6*: 693–700, 2000.

53. Kreitman, R.J., Bailon, P., Chaudhary, V.K., FitzGerald, D.J.P., and Pastan, I. Recombinant immunotoxins containing anti-Tac(Fv) and derivatives of *Pseudomonas* exotoxin produce complete regression in mice of an interleukin-2 receptor-expressing human carcinoma. *Blood, 83*: 426–434, 1994.

54. Kreitman, R.J. and Pastan, I. Targeting *Pseudomonas* exotoxin to hematologic malignancies. *Semin. Cancer Biol.*, 6: 297–306, 1995.

55. Kreitman, R.J., Chaudhary, V.K., Kozak, R.W., FitzGerald, D.J.P., Waldmann, T.A., and Pastan, I. Recombinant toxins containing the variable domains of the anti-Tac monoclonal antibody to the interleukin-2 receptor kill malignant cells from patients with chronic lymphocytic leukemia. *Blood*, 80: 2344–2352, 1992.

56. Kreitman, R.J., Chang, C.N., Hudson, D.V., Queen, C., Bailon, P., and Pastan, I. Anti-Tac(Fab)-PE40, a recombinant double-chain immunotoxin which kills interleukin-2-receptor-bearing cells and induces complete remission in an *in vivo* tumor model. *Int. J. Cancer*, 57: 856–864, 1994.

57. Kreitman, R.J. and Pastan, I. Accumulation of a recombinant immunotoxin *in vivo*: less than 1000 molecules per cell is sufficient for complete responses. *Cancer Res.*, 58: 968–975, 1998.

58. Kreitman, R.J. Quantification of immunotoxin number for complete therapeutic response. *Methods Mol. Biol.*, 166: 111–123, 2000.

59. Kobayashi, H., Kao, C.K., Kreitman, R.J., Le, N., Kim, M., Brechbiel, M.W., Paik, C.H., Pastan, I., and Carrasquillo, J.A. Pharmacokinetics of In-111- and I-125-labeled antiTac single-chain Fv recombinant immunotoxin. *J. Nuc. Med.*, 41: 755–762, 2000.

60. Kreitman, R.J., Wilson, W.H., Robbins, D., Margulies, I., Stetler-Stevenson, M., Waldmann, T.A., and Pastan, I. Responses in refractory hairy cell leukemia to a recombinant immunotoxin. *Blood*, 94: 3340–3348, 1999.

61. Kreitman, R.J., Wilson, W.H., White, J.D., Stetler-Stevenson, M., Jaffe, E.S., Waldmann, T.A., and Pastan, I. Phase I trial of recombinant immunotoxin Anti-Tac(Fv)-PE38 (LMB-2) in patients with hematologic malignancies. *J. Clin. Oncol.*, 18: 1614–1636, 2000.

62. Onda, M., Kreitman, R.J., Vasmatzis, G., Lee, B., and Pastan, I. Reduction of the nonspecific toxicity of anti-Tac(Fv)-PE38 by mutations in the framework regions of the Fv which lower the isoelectric point. *J. Immunol.*, 163: 6072–6077, 1999.

63. Onda, M., Willingham, M., Wang, Q., Kreitman, R.J., Tsutsumi, Y., Nagata, S., Dinarello, C.A., and Pastan, I. Inhibition of TNF alpha produced by Kupffer cells protects against the non-specific liver toxicity of immunotoxin anti-Tac(Fv)-PE38, LMB-2. *J. Immunol.*, 165: 7150–7156, 2000.

64. Mavroudis, D.A., Jiang, Y.Z., Hensel, N., Lewalle, P., Couriel, D., Kreitman, R.J., Pastan, I., and Barrett, A.J. Specific depletion of alloreactivity against haplotype mismatched related individuals: a new approach to graft-versus-host disease prophylaxis in haploidentical bone marrow transplantation. *Bone Marrow Transplant.*, 17: 793–799, 1996.

65. Harris, D.T., Sakiestewa, D., Lyons, C., Kreitman, R.J., and Pastan, I. Prevention of graft-versus-host disease (GVHD) by elimination of recipient-reactive donor T cells with recombinant toxins that target the interleukin 2 (IL-2) receptor. *Bone Marrow Transplant.*, 23: 137–144, 1999.

66. Reiter, Y., Brinkmann, U., Kreitman, R.J., Jung, S.-H., Lee, B., and Pastan, I. Stabilization of the Fv fragments in recombinant immunotoxins by disulfide bonds engineered into conserved framework regions. *Biochemistry*, 33: 5451–5459, 1994.

67. Reiter, Y., Kreitman, R.J., Brinkmann, U., and Pastan, I. Cytotoxic and antitumor activity of a recombinant immunotoxin composed of disulfide-stablized anti-Tac Fv fragment and truncated *Pseudomonas* exotoxin. *Int. J. Cancer*, 58: 142–149, 1994.

68. Onda, M., Nagata, S., Tsutsumi, Y., Vincent, J.J., Wang, Q.C., Kreitman, R.J., Lee, B., and Pastan, I. Lowering the isoelectric point of the Fv portion of recombinant immunotoxins leads to decreased nonspecific animal toxicity without affecting antitumor activity. *Cancer Res., 61*: 5070–5077, 2001.

69. Tsutsumi, Y., Onda, M., Nagata, S., Lee, B., Kreitman, R.J., and Pastan, I. Site-specific chemical modification with polyethylene glycol of recombinant immunotoxin anti-Tac(Fv)-PE38 (LMB-2) improves antitumor activity and reduces animal toxicity and immunogenicity. *Proc. Natl. Acad. Sci. USA, 97*: 8548–8553, 2000.

70. Barth, S., Huhn, M., Wels, W., Diehl, V., and Engert, A. Construction and *in vitro* evaluation of RFT5(scFv)-ETA', a new recombinant single-chain immunotoxin with specific cytotoxicity toward CD25+ Hodgkin-derived cell lines. *Int. J. Molec. Med., 1*: 249–256, 1998.

71. Barth, S., Huhn, M., Matthey, B., Schnell, R., Tawadros, S., Schinkothe, T., Lorenzen, J., Diehl, V., and Engert, A. Recombinant anti-CD25 immunotoxin RFT5(ScFv)-ETA' demonstrates successful elimination of disseminated human Hodgkin lymphoma in SCID mice. *Int. J. Cancer, 86*: 718–724, 2000.

72. Law, C.L., Aruffo, A., Chandran, K.A., Doty, R.T., and Clark, E.A. Ig domains 1 and 2 of murine CD22 constitute the ligand-binding domain and bind multiple sialylated ligands expressed on B and T cells. *J. Immunol., 155*: 3368–3376, 1995.

73. Otipoby, K.L., Andersson, K.B., Draves, K.E., Klaus, S.J., Farr, A.G., Kerner, J.D., Perlmutter, R.M., Law, C.L., and Clark, E.A. CD22 regulates thymus-independent responses and the lifespan of B cells. *Nature, 384*: 634–637, 1996.

74. Kreitman, R.J., Hansen, H.J., Jones, A.L., FitzGerald, D.J.P., Goldenberg, D.M., and Pastan, I. *Pseudomonas* exotoxin-based immunotoxins containing the antibody LL2 or LL2-Fab' induce regression of subcutaneous human B-cell lymphoma in mice. *Cancer Res., 53*: 819–825, 1993.

75. Shen, G., Li, J., Ghetie, M., Ghetie, V., May, R.D., Till, M., Brown, A.N., Relf, M., Knowles, P., Uhr, J.W., Janossy, G., Amlot, P., Vitetta, E.S., and Thorpe, P.E. Evaluation of four CD22 antibodies as ricin A chain-containing immunotoxins for the *in vivo* therapy of human B-cell leukemias and lymphomas. *Int. J. Cancer, 42*: 792–797, 1988.

76. Ghetie, M.-A., May, R.D., Till, M., Uhr, J.W., Ghetie, V., Knowles, P.P., Relf, M., Brown, A., Wallace, P.M., Janossy, G., Amlot, P., Vitetta, E.S., and Thorpe, P.E. Evaluation of ricin A chain-containing immunotoxins directed against CD19 and CD22 antigens on normal and malignant human B-Cells as potential reagents for *in vivo* therapy. *Cancer Res., 48*: 2610–2617, 1988.

77. Ghetie, M.-A., Tucker, K., Richardson, J., Uhr, J.W., and Vitetta, E.S. The antitumor activity of an anti-CD22 immunotoxin in SCID mice with disseminated Daudi lymphoma is enhanced by either an anti-CD19 antibody or an anti-CD19 immunotoxin. *Blood, 80*: 2315–2320, 1992.

78. Amlot, P.L., Stone, M.J., Cunningham, D., Fay, J., Newman, J., Collins, R., May, R., McCarthy, M., Richardson, J., Ghetie, V., Ramilo, O., Thorpe, P.E., Uhr, J.W., and Vitetta, E.S. A phase I study of an anti-CD22-deglycosylated ricin A chain immunotoxin in the treatment of B-cell lymphomas resistant to conventional therapy. *Blood, 82*: 2624–2633, 1993.

79. Vitetta, E.S., Stone, M., Amlot, P., Fay, J., May, R., Till, M., Newman, J., Clark, P., Collins, R., Cunningham, D., Ghetie, V., Uhr, J., and Thorpe, P.E. Phase I immunotoxin trial in patients with B-cell lymphoma. *Cancer Res., 51*: 4052–4058, 1991.

80. Mansfield, E., Pastan, I., and FitzGerald, D.J. Characterization of RFB4-*Pseudomonas* exotoxin A immunotoxins targeted to CD22 on B-cell malignancies. *Bioconjugate Chem.*, 7: 557–563, 1996.

81. Mansfield, E., Chiron, M.F., Amlot, P., Pastan, I., and FitzGerald, D.J. Recombinant RFB4 single-chain immunotoxin that is cytotoxic towards CD22-positive cells. *Biochem. Soc. Trans.*, 25: 709–714, 1997.

82. Mansfield, E., Amlot, P., Pastan, I., and FitzGerald, D.J. Recombinant RFB4 immunotoxins exhibit potent cytotoxic activity for CD22-bearing cells and tumors. *Blood*, 90: 2020–2026, 1997.

83. Kreitman, R.J., Wang, Q.C., FitzGerald, D.J.P., and Pastan, I. Complete regression of human B-cell lymphoma xenografts in mice treated with recombinant anti-CD22 immunotoxin RFB4(dsFv)-PE38 at doses tolerated by Cynomolgus monkeys. *Int. J. Cancer*, 81: 148–155, 1999.

84. Kreitman, R.J., Margulies, I., Stetler-Stevenson, M., Wang, Q.C., FitzGerald, D.J.P., and Pastan, I. Cytotoxic activity of disulfide-stabilized recombinant immunotoxin RFB4(dsFv)-PE38 (BL22) towards fresh malignant cells from patients with B-cell leukemias. *Clin. Cancer Res.*, 6: 1476–1487, 2000.

85. Kreitman, R.J., Wilson, W.H., Bergeron, K., Raggio, M., Stetler-Stevenson, M., FitzGerald, D.J., and Pastan, I. Efficacy of the Anti-CD22 Recombinant Immunotoxin BL22 in Chemotherapy-Resistant Hairy-Cell Leukemia. *New. Engl. J. Med.*, 345: 241–247, 2001.

86. Piro, L.D., Carrera, C.J., Carson, D.A., and Beutler, E. Lasting remissions in hairy-cell leukemia induced by a single infusion of 2-chlorodeoxyadenosine. *N. Engl. J. Med.*, 322: 1117–1121, 1990.

87. Saven, A., Burian, C., Koziol, J.A., and Piro, L.D. Long-term follow-up of patients with hairy cell leukemia after cladribine treatment. *Blood*, 92: 1918–1926, 1998.

88. Cheson, B.D., Sorensen, J.M., Vena, D.A., Montello, M.J., Barrett, J.A., Damasio, E., Tallman, M., Annino, L., Connors, J., Coiffier, B., and Lauria, F. Treatment of hairy cell leukemia with 2-chlorodeoxyadenosine via the Group C protocol mechanism of the National Cancer Institute: a report of 979 patients. *J. Clin. Oncol.*, 16: 3007–3015, 1998.

89. Goodman, G.R., Burian, C., Koziol, J.A., and Saven, A. Extended follow-up of patients with hairy cell leukemia after treatment with cladribine. *J. Clin. Oncol.*, 21: 891–896., 2003.

90. Tallman, M.S., Hakimian, D., Kopecky, K.J., Wheaton, S., Wollins, E., Foucar, K., Cassileth, P.A., Habermann, T., Grever, M., Rowe, J.M., and Peterson, L.C. Minimal residual disease in patients with hairy cell leukemia in complete remission treated with 2-chlorodeoxyadenosine or 2-deoxycoformycin and prediction of early relapse. *Clin. Cancer. Res.*, 5: 1665–1670, 1999.

91. Chowdhury, P.S. and Pastan, I. Improving antibody affinity by mimicking somatic hypermutation *in vitro*. *Nat. Biotechnol.*, 17: 568–572, 1999.

92. Neuberger, M.S. and Milstein, C. Somatic hypermutation. *Curr. Opin. Immunol.*, 7: 248–254, 1995.

93. Salvatore, G., Beers, R., Kreitman, R.J., and Pastan, I. Improved Cytotoxic activity towards cell lines and fresh leukemia cells of a mutant anti-CD22 immunotoxin obtained by antibody phage display. *Cancer Res.*, 8: 995–1002, 2002.

94. Dürkop, H., Latza, U., Hummel, M., Eitelbach, F., Seed, B., and Stein, H. Molecular cloning and expression of a new member of the nerve growth factor receptor family that is characteristic for Hodgkin's disease. *Cell*, 68: 421–427, 1992.

95. Kadin, M.E. and Morris, S.W. The t(2;5) in human lymphomas. *Leuk. Lymphoma*, *29*: 249, 1998.
96. Higgins, J.P. and Warnke, R.A. CD30 expression is common in mediastinal large B-cell lymphoma [see comments]. *Am. J. Clin. Pathol.*, *112*: 241–247, 1999.
97. Rozemuller, H., Chowdhury, P.S., Pastan, I., and Kreitman, R.J. Isolation of new anti-CD30 scFvs from DNA immunized mice by phage display and biologic activity of recombinant immunotoxins produced by fusion with truncated Pseudomonas exotoxin. *Int. J. Cancer.*, *92*: 861–870, 2001.
98. Nagata, S., Onda, M., Numata, Y., Santora, K., Beers, R., Kreitman, R.J., and Pastan, I. Novel Anti-CD30 recombinant immunotoxins containing disulfide-stabilized Fv fragments. *Clin. Cancer. Res.*, *8*: 2345–2355., 2002.
99. Ogata, M., Fryling, C.M., Pastan, I., and FitzGerald, D.J. Cell-mediated cleavage of *Pseudomonas* exotoxin between Arg^{279} and Gly^{280} generates the enzymatically active fragment which translocates to the cytosol. *J. Biol. Chem.*, *267*: 25396–25401, 1992.

chapter two

Interleukin-4-targeted cytotoxin for recurrent glioblastoma therapy

Koji Kawakami, Mariko Kawakami, Robert J. Kreitman, and Raj K. Puri

Contents

Previous studies have reported that receptors for interleukin-4 (IL-4), a pleiotropic immunoregulatory cytokine, are overexpressed in a variety of human solid cancer cell lines and primary cell cultures. As IL-4 can upregulate adhesion molecules, inhibit cell proliferation, and mediate signal transduction in tumor cell lines, these receptors are considered functional. To target IL-4R, we have developed a chimeric fusion protein composed of a circularly permuted IL-4 and a mutated form of *Pseudomonas* exotoxin [termed IL4(38-37)-PE38KDEL or cpIL4-PE]. Recombinant cpIL4-PE was highly cytotoxic to cancer cells, while not cytotoxic or less cytotoxic to human hematopoietic cells

0-4152-6365-4/05/$0.00+$1.50
© 2005 by CRC Press

including immune cells, bone marrow derived CD34$^+$ cells, and normal human astrocytes. This agent showed remarkable antitumor activity in glioblastoma multiforme, prostate cancer, lung cancer, pancreatic cancer, breast cancer, AIDS-Kaposi's sarcoma, and head and neck cancer models in immunodeficient animals. cpIL4-PE caused partial or complete regression of established human tumors and responses were very durable. These and other encouraging preclinical efficacy, safety, and tolerability studies lead to the testing of cpIL4-PE in patients with recurrent glioblastoma. Initially, cpIL4-PE was administered directly into the brain tumors of nine patients. An extended phase I/II clinical trial was recently completed to determine the safety, tolerability, and efficacy of cpIL4-PE in patients with recurrent glioma, whose tumors were directly injected stereotactically using convection enhanced delivery (CED). Our preliminary clinical results suggest that cpIL4-PE can cause pronounced necrosis of recurrent glioma tumors without systemic toxicity. The CNS toxicities observed were attributed to the volume of infusion and/or nonspecific toxicity. Additional clinical trials will reveal the antitumor activities of IL-4 cytotoxin in recurrent malignant glioma.

Introduction

Interleukin-4 (IL-4) is predominantly produced by activated T lymphocytes, mast cells, and basophils, and has been shown to mediate many effects on numerous cell types including T cells, B cells, monocytes, mast cells, endothelial cells, fibroblasts, astrocytes, and osteoblasts.[1-3] IL-4 has also been shown to have direct modest growth inhibitory effects on hematopoietic and non-hematopoietic tumor cell lines *in vitro* and *in vivo*.[4-9] These results brought IL-4 into clinical trials for the treatment of hematopoietic and non-hematopoietic malignancies.[10-12] Because the clinical results were disappointing, it appears that further clinical studies using IL-4 as an anticancer agent are not being conducted.

We and others have reported that receptors for IL-4 (IL-4R) are overexpressed on a variety of hematopoietic and solid tumor cell including glioblastoma.[2,3,13-15] Although the significance of the overexpression of IL-4R on tumor cell lines is unknown, we have produced a recombinant fusion protein termed cpIL4-PE that targets IL-4R on tumor cells.[14] In this chapter, we summarize the expression and structure of IL-4R on solid tumor cell lines and discuss our preclinical and preliminary clinical results using cpIL4-PE for the treatment of recurrent glioblastoma multiforme.

Expression and structure of IL-4R on human brain tumor cell lines

Normal cells including T cells, B cells, monocytes, eosinophils, basophils, fibroblasts, and endothelial cells express low numbers of IL-4R, which are sufficient for IL-4 to induce biological functions in these cells.[16-21] On the other hand, we have reported that a variety of human solid tumor cell lines express moderate to high numbers of IL-4R.[3,13–15,22–35] In radiolabeled IL-4 binding studies, tumor cell lines were found to express intermediate to high affinity IL-4R (Table 2.1). To demonstrate expression of IL-4R *in situ*, surgical samples of high-grade astrocytoma and glioblastoma were assessed for the expression of IL-4R by reverse transcriptase polymerase chain reaction (RT-PCR) and Southern blot analysis. We demonstrated that 16 of 21 surgical samples expressed IL-4Rα chain.[28] In addition, 25/25 samples from another series of primary brain tumor primary cell cultures expressed mRNA for IL-4R as assessed by RT-PCR.[34] Out of 32 brain tumor samples, 18 were obtained from glioblastoma multiforme (GBM) and 14 samples represented other central nervous system tumors. Fifteen of eighteen (83%) GBM tumors and six of seven (86%) astrocytoma were found to be moderately to highly positive for IL-4R expression *in situ*.[35] Although in our first study, six normal brain tissues (five from the frontal lobe and one from the temporal of the cortex) obtained from six individuals did not express detectable mRNA for IL-4Rα chain, later studies have shown that six additional normal brain tissues did express IL-4Rα mRNA by RT-PCR analysis.[28] However, a recent study did not show detectable IL-4R protein in normal brain tissues by immunohistochemistry even though IL-4Rα chain mRNA was expressed.[35]

Table 2.1 IL-4R expression in human normal cells and tumor cell lines

Cell type	IL-4 binding sites/cell	Kd (pM)[a]	Reference
Normal cells			
Basophils	300–600	70–100	16
Endothelial cells	<50		22
Resting T and B cells	<500	25–100	17–19
Monocytes	200–300	25–100	20, 21
Tumor cell lines			
Renal cell carcinoma	1400–4000	100–300	15, 23
Malignant melanoma	1200–1400	360–550	24
Ovarian cancer	300–1400	330	24
Kaposi's sarcoma	600–2200	19–160	22, 25, 94
Epidermoid carcinoma	1000	300	Unpublished data
Breast cancer	700–4600	200–1000	24, 26
Glioblastoma	1000–3000	100–700	27
Colon cancer	2000	77	4
Pancreatic cancer	9200	370	31
Head and neck cancer	6100–13,000	—	29, 30, 105
Prostate cancer	12,000	266	33
Gastric cancer	—[b]	—	7
Lung cancer	10,600	2400	32, 61

[a] Kd, dissociation constant.
[b] Flow cytometric analysis.

Table 2.2 Expression of IL-4Rα mRNA and proteins *in situ*

	Ref. 28	Ref. 34	Ref. 35
Tissues	RT-PCR[a]	RT-PCR	IHC[b]
Glioblastoma	15/19	17/17	15/18
Malignant astrocytoma	1/2	2/2	6/7
Oligodendroglioma	—	3/3	3/3
Normal brain	1/6	2/2	0/4

[a] RT-PCR, reverse transcriptase-polymerase chain reaction.
[b] IHC, immunohistochemistry.

These results suggested that IL-4Rs are differentially overexpressed in glioma samples compared to the normal brain (Table 2.2). Additional samples need to be further evaluated to confirm these findings and perhaps define the heterogeneity of pathologically confirmed glioblastoma multiforme tumors.

A major component of the IL-4R is a 140 kDa protein originally termed IL-4Rα.[36] It contains a WSXWS motif and four cysteine residues at a fixed location and long intracellular domain.[37] A second subunit of the IL-4R system was identified as a component of the IL-2 receptor system, the γ chain.[38,39] IL-2 receptor γ chain forms a functional complex with the IL-4Rα chain. Later the IL-2Rγ chain was also shown to be a component of the IL-7, IL-9, IL-15, and IL-21 receptor systems.[40–43]

To determine the subunit structure of the IL-4 receptor on human solid tumor cell lines, affinity cross-linking, Northern analysis, RT-PCR, and immunoprecipitation studies were performed.[30,44,45] Human renal cell carcinoma, glioblastoma, colon carcinoma, ovarian carcinoma and melanoma cell lines expressed many high affinity IL-4R sites and these receptors were shown to consist of 140 kDa and 70 kDa proteins.[15,23,24,45] However, unlike immune cells, antibody to γ_c did not immunoprecipitate any bands indicating that γ_c did not participate in the formation of the IL-4R complex. The lack of γ_c chain was further confirmed by Northern analysis.[44] Since a 70 kDa protein was identified in addition to the IL-4Rα chain in tumor cell lines, we discovered a new chain for the IL-4R complex. This protein was later cloned in the IL-13R system and termed IL-13Rα1 chain.[46–49] To determine the interaction between IL-4 and IL-13 receptor systems, cross-competition studies were performed using IL-13. We determined whether binding of ^{125}I-IL-4 was competed by an excess of IL-13. In most cell lines tested, unlabeled IL-13 inhibited IL-4 binding to its receptors.[24,30,44,50] These results and other studies suggested that in human solid tumor cells, IL-4R and IL-13R are related and may share chain(s) with each other.[45,50–56] Reconstitution studies suggested that IL-13Rα1 chain can form a complex with IL-4Rα chain in the formation of a functional IL-4R complex.[51,57] These studies further demonstrated that the IL-13Rα1 chain can substitute for γ_c in mediating IL-4 signaling and thus this chain forms a third subunit of the IL-4R system (IL-4Rα, IL-2Rγ_c, and IL-13Rα1).[51,56–59] Based on these studies, we proposed that immune T cells

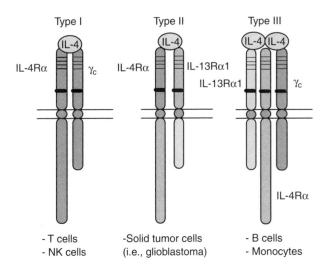

Figure 2.1 Model of IL-4R

and NK cells express type I IL-4R and tumor cells express type II IL-4R (Figure 2.1). However, there are certain immune cells (e.g., B cells and monocytes) that express type III IL-4R. Whether all three chains form a functional complex with IL-4 is not known.

IL-4R on solid tumor cells are considered to be functional because IL-4 is able to mediate signal transduction, inhibit cell growth, and increase expression of major histocompatibility antigens and intracellular adhesion molecule-1 (ICAM-1) in some tumor cell lines.[60–64] IL-4 utilizes JAK/STAT pathways for signaling in both immune and solid tumor cell lines.[51,57] In immune cells, IL-4 phosphorylates and activates JAK1 and JAK3 tyrosine kinases, however, in certain tumor cell lines it phosphorylates and activates JAK1 and JAK2 tyrosine kinases.[45,51] Irrespective of these differences IL-4 activates STAT6 (signal transduction and activators of transcription) in both cell types.[51]

Although growth–inhibitory effects of IL-4 were seen in some solid tumor cell lines, IL-4 did not modulate tumor growth in glioblastoma cells.[27] The reason for this difference is unclear. Glioma cells expressed a similar configuration of IL-4R as seen in renal cell carcinoma cells, in which IL-4 can inhibit proliferation *in vitro*.[15] Recently, the IL-13Rα2 chain[65,66] has been shown to act as a decoy receptor for IL-13 and inhibit IL-4-dependent signal transduction in glioblastoma cells.[67–72] Additional studies are needed to define the significance of IL-4R expression in glioma cells.

Development of IL-4R-targeted cytotoxin (in vitro *studies*)

To target the IL-4R that were found to be overexpressed on solid tumor cells, we generated a chimeric protein composed of IL-4 and a truncated form of *Pseudomonas* exotoxin, IL-4 cytotoxin. This form of fusion protein, called an immunotoxin or cytotoxin, targets specific cell-surface antigens or receptors on cancer cells.[73–79] *Pseudomonas* exotoxins (PE) is a natural bacterial toxin secreted by *Pseudomonas aeruginosa*. PE has three domains: domain Ia binds to PE receptors, domain II catalyzes the translocation of the toxin into the cytosol, and domain III inhibits protein synthesis by inhibition of ADP ribosylation of elongation factor 2.[75] The receptors for PE are ubiquitously expressed on many types of cells, and cells are killed after binding to PE. However, by replacing the binding domain of PE with human IL-4, one can subsequently decrease nonspecific cytotoxic activity and assign a new specific function to PE. This fusion protein will only bind to cells expressing IL-4R on their cell surface, and these cells will be killed only when they are able to internalize PE molecules that are processed in the correct intracellular compartment of the cytoplasm. This chimeric toxin, IL4-PE[4E], was first developed by the fusion of human IL-4 cDNA to the 5' end of cDNA encoding a full-length PE containing mutations in domain Ia of the toxin.[14,79–81] This toxin was produced by expressing the chimeric gene in *Escherichia coli*, and then purifying the recombinant monomer from inclusion bodies. The purified IL4- PE[4E] was found to be highly cytotoxic to brain tumor cell lines, while irrelevant fusion toxins IL2-PE[4E] or IL6-PE[4E] were not cytotoxic to brain tumor cells.[27] The cytotoxicity of IL4-PE[4E] was specific because it was neutralized by an excess of human IL-4 in all brain tumor cell lines tested. IL4-PE[4E] was not cytotoxic to many types of normal human cells.[28] In addition, an IL-4 cytotoxin mutant lacking ADP-ribosylating activity in domain III of PE (IL4-PE[4D]) was not cytotoxic to brain tumor cells, indicating that the IL-4 cytotoxin-mediated cytotoxicity of the tumor cells required both IL-4R binding and enzymatic toxin activity.[27]

Because IL4-PE[4E] bound to IL-4R in human glioma cells with 37-fold less binding affinity than IL-4 itself, we produced a circularly permuted IL-4 cytotoxin, termed IL4(38-37)-PE38KDEL or cpIL4-PE as shown in Figure 2.2.[82–84] This molecule contains amino acids 38–129 of IL-4, fused via a peptide linker to amino acids 1–37, which in turn is fused to amino acids 353–364 and 381–608 of PE, with KDEL (an endoplasmic retaining sequence),

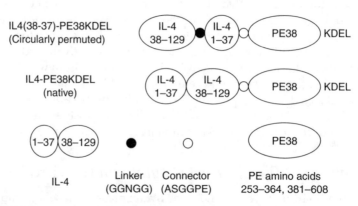

Figure 2.2 Schematic model of IL-4 cytotoxins

at positions 609–612. This purified cytotoxin (cpIL4-PE) was found to be highly cytotoxic to IL-4R-positive human solid cancer cell lines. Recombinant cpIL4-PE bound to human glioblastoma cells with 16-fold higher affinity than the native noncircular permuted IL-4 cytotoxins, IL4-PE[4E], or IL4-PE38KDEL.[82,83] In addition, cpIL-4PE was 3- to 30-fold more cytotoxic to glioblastoma cell lines compared with the first-generation toxins IL4-PE[4E] or noncircular permuted IL4-PE38KDEL.[28] As shown in Table 2.3, the IC_{50} representing the concentration of IL-4 cytotoxin, cpIL4-PE, at which 50% inhibition of protein synthesis is observed compared with untreated cells, was around 10 ng/ml in most (17 of 23) brain tumor cell lines.[13,28,35] The cytotoxic activity of cpIL4-PE was specific, since excess IL-4 neutralized its cytotoxic activity to the cancer cells. In addition, although the cell death mechanism of cpIL4-PE to each individual cancer cell type has not been fully

Table 2.3 Cytotoxic activity of cpIL4-PE to glioma cell lines and primary cell cultures

	IC_{50} (ng/ml)
Glioblastoma, glioma cell lines **(Refs. 3, 13)**	
T98G	6
A172	5
U87MG	50
U373MG	12
SN19	20
SF295	150
U251	15
H638	25
Glioblastoma explants **(Ref. 35)**	
GBM1	200
GBM2	10
GBM3	5
GBM4	10
GBM5	12
GBM6	12
GBM7	750
GBM8	13
GBM9	12
GBM10	12
GBM11	5
GBM12	4
GBM13	1
GBM14	2
GBM15	6
Normal brain cell line	
Normal human astrocyte (NHA)	350
Human neuronal cell (NT-2)	>1000

investigated, we and others have reported that PE-based cytotoxins can cause apoptotic cell death in addition to necrosis.[30,85–89]

The potent cytotoxic activity of cpIL-4PE was confirmed by clonogenic assays.[30,90,91] Tumor colonies formed by U251 cells were sensitive to cpIL4-PE in a dose-dependent manner at levels similar to those observed in protein synthesis inhibition assays. More than 50% of colonies were inhibited at 1 ng/ml of cpIL4-PE.[90] In addition, the growth of two glioblastoma cell lines, U251 and T98G, was inhibited in a time-dependent manner. More than 91% of U251 colonies were killed within 4 h, and 86% of T98G colonies were killed within 24 h. Continuous exposure to IL-4 cytotoxin for 10 days completely inhibited colony formation in both cell lines.[90] Because clonogenicity *in vitro* correlates with *in vivo* malignant phenotype in xenografts,[92,93] our studies suggest that cpIL4-PE will have antitumor activity *in vivo*.

In vivo *antitumor activity of IL-4 cytotoxin in animal models of glioblastoma*

Based on the exciting *in vitro* results, the *in vivo* antitumor activity of cpIL4-PE was investigated. Athymic nude mice were injected subcutaneously (s.c.) with human cancer cells including glioblastoma, pancreatic cancer, prostate cancer, head and neck cancer, lung cancer, breast cancer, and AIDS Kaposi's sarcoma.[26,31–33,83,90,94,95] When U251 glioblastoma cells were injected s.c. in nude mice, tumor developed to a mean size of 13 to 30 mm^2 after 3 to 6 days.[90] These cells consistently generated solid tumors in all injected animals. The efficacy of cpIL4-PE when administered by different routes and dosing schedules was evaluated.

Intraperitoneal (i.p.) administration of cpIL4-PE on alternate days for a total of three injections significantly inhibited tumor growth in a dose-dependent manner. Complete tumor regressions were seen at the two highest i.p. doses (50 or 100 μg/kg) in U251 s.c. established tumors. Three of five U251 tumor-bearing mice were complete responders in the highest dose treated groups, and remained tumor-free during a long-term follow-up period. In mice treated with the lowest dose (25 μg/kg), the size of tumors was significantly smaller than mice treated with excipient alone (control). All animals tolerated the therapy well without any evidence of nonspecific toxicity.

Antitumor activity of cpIL4-PE was also assessed by intravenous (i.v.) administration. U251 tumor bearing mice were injected i.v. with one of three different doses of cpIL4-PE on alternate days for a total of three injections and then tumor sizes were measured. The three different doses selected were 50, 100, or 200 μg/kg. Tumor regression was noted in all of the mice treated with cpIL4-PE at all dose levels. The effect of cpIL4-PE was clearly dose dependent; however, complete regressions were not observed. This was attributed to short serum half-life and drug availability at the tumor site. Improved delivery methods or perhaps continuous infusions would lead to complete responses as seen in the i.p. route of cpIL-4PE administration.[96]

Because glioma is a localized central nervous system disease, it is possible to administer therapeutic agents directly into the tumor bed. Therefore, we investigated whether intratumoral (i.t.) treatment would lead to improved antitumor activity. U251 s.c. tumor bearing nude mice were injected i.t. with various doses and schedules of cpIL4-PE (up to 1000 μg/kg). When tumors were injected with 250 μg/kg of cpIL4-PE (alternate days for three injections), they began to decrease significantly in size by day 10, and all of the treated mice exhibited complete regression by day 24. Although tumors recurred in 50% of the mice by day 37 from tumor implantation, the mean size of the tumors remained significantly smaller than that of the control group (2% of control tumor size) through day 58. Based on these promising results, we evaluated the antitumor activity of cpIL4-PE against relatively larger (60 mm²) U251 s.c. established tumors. Complete tumor regression in 100% of the mice treated with 750 μg/kg cpIL4-PE ×3 days on alternate days was seen and mice remained tumor-free for at least 3 months. Taken together, these results suggested that cpIL-4PE can mediate remarkable anti-glioma activity, which encouraged us to consider the initiation of a Phase I clinical trial in patients with recurrent glioma.

Toxicology and pharmacokinetics of IL-4 cytotoxin in preclinical studies

In vivo toxicology and pharmacokinetic studies in mice, rats, guinea pigs, and cynomolgus monkeys were conducted.[83,94] The LD_{50} in mice was determined to be 475 μg/kg administered every other day for a total of three injections. At 400 μg/kg and higher doses administered every other day for a total of three injections, liver enzymes were elevated. Thus, the dose-limiting toxicity (DLT) in mice was liver toxicity. This toxicity was most likely related to nonspecific uptake of the toxin by the liver since human IL-4 does not bind murine cells.[97] The serum half life ($t_{1/2}$) of cpIL4-PE was 10 min after a single intravenous administration of the drug in nude mice.[83]

To determine the concentration of cpIL4-PE required to cause necrosis in normal brain tissues by nonspecific internalization into cells, we employed a rat model. Rat IL-4R also do not bind human IL-4. Six groups of rats were injected in the frontal cortex with various doses of cpIL4-PE. On day 4, the animals were sacrificed and their brain tissues were examined microscopically. All animals survived until the terminal necropsy. No abnormalities in behavior were noted and the mean body weights did not change in all groups. There were no remarkable gross pathology findings. No histopathological changes were observed at cpIL4-PE concentrations of <100 μg/kg. However, microscopic evaluation revealed necrosis of the right cortical hemisphere at the injection site in the group receiving 1000 μg/kg cpIL4-PE.[28]

We also performed intrathecal administration of cpIL4-PE in monkeys as a pharmacologic/toxicologic model for intratumor injection of patients. Because human IL-4 is primate-specific and can bind to monkey fibroblast and gibbon ape leukemia cells, this model allowed us to evaluate the pharmacologic and toxicologic consequences of cpIL4-PE injection into the normal brain.[41,98,99] To determine the CSF levels of cpIL4-PE after intrathecal administration, cynomolgus monkeys were injected with 0.2% human serum albumin-PBS containing 0, 2, or 6 μg/kg of cpIL4-PE on days 1, 3, and 5.[28] The CSF samples were drawn at various time points. Serum samples were also collected 2 h after injection on days 3 and 5 at each dose level. High cpIL4-PE levels were measurable in CSF after administration of the drug. Because the CSF volume is approximately 1 ml/kg, the levels at 2 h were >15% of peak values expected if the drug were to immediately distribute throughout the CSF. cpIL4-PE was cleared rapidly from the CSF, as at 24 h after injection less than 1% of the drug remained. No detectable serum levels of cpIL4-PE were observed in any group. The same monkeys utilized for pharmacokinetic studies were also evaluated for any signs or alterations in hematology and clinical chemistry alterations as an indicator of toxicity. No systemic toxicity was observed as expected by the absence of the drug in the serum. No changes in hematology and serum chemistry were observed in any of the monkeys studied, consistent with the absence of cpIL4-PE in the serum. Creatinine phosphokinase (CPK) BB bands, which can be a sensitive enzyme indicator for brain injury, were not elevated in any of the monkeys examined.

Cynomolgus monkeys were also injected intravenously with two doses of cpIL4-PE and serum chemistry and hematological parameters were assessed. These monkeys were administered 50 and 200 μg/kg doses, on alternate days for three injections. Both monkeys tolerated these doses well, however, reversible hepatic toxicities were observed. High-peak serum levels of the drug were observed at both doses (7 and 2.5 μg/ml at 200 and 50 μg/kg doses) (Kreitman, R.J., Puri, R.K., and Pastan, I., unpublished results). These studies suggested that high-peak concentrations of cpIL4-PE are well tolerated without major toxicities.

Interestingly, only ~10 to 100 ng/ml of cpIL4-PE was sufficient to kill more than 95% of glioblastoma cells in tissue culture. However, a 17-fold higher concentration of cpIL4-PE at the 6 μg/kg dose and a ~3- to 4-fold higher concentration at the 2 μg/kg dose by intrathecal route did not cause any noticeable pathological changes.[28] These results therefore predict that sufficient levels of IL-4 cytotoxin could be achieved for therapeutic efficacy by administration of only 1 to 2 μg/kg of cpIL4-PE. Higher concentrations of cpIL4-PE may not be needed and thus will avoid nonspecific toxicity.

Clinical trials with IL-4 cytotoxin in recurrent glioblastoma patients

Based on our preclinical results, we initiated a Phase I clinical trial to determine the safety and tolerability of cpIL4-PE in recurrent human malignant glioblastoma when injected intratumorally by CED utilizing two to three catheters.[100] This trial was initiated at the John Wayne Cancer Institute and

St. John's Hospital, Santa Monica, California. Our protocol was approved by the Institutional Review Boards of the Food and Drug Administration (FDA) and St. John's Hospital, and an Investigational New Drug (IND) application was allowed by the FDA. cpIL4-PE was infused by Dr. Robert Rand of the John Wayne Cancer Institute. The cpIL4-PE was produced following current good manufacturing practices (cGMP).

Nine patients with recurrent brain tumors who failed standard therapy including radiation therapy and/or chemotherapy were enrolled. These patients did not receive any other treatment within 3 weeks of inclusion into the trial. All nine patients had adequate baseline organ functions, as assessed by standard laboratory parameters on their preoperation visit. Patients with diffuse tumors or cerebrospinal fluid metastasis were excluded from the study.

All patients underwent a standard stereotactic biopsy under CT, as previously described.[101] Up to three silastic infusion catheters (2.1 mm outer diameter, Pudenz catheter[s]) were placed with the tip at a selected site in the tumor using stereotactic guidance through small twist drill holes. The sites for catheter placement were judged by individual neurosurgeons. The numbers of catheters were selected on the basis of the volume to be infused and to ensure maximum saturation of the tumor bed and margins in the designated period of time. After surgery, the externalized catheters were connected to Medex 2010 micropumps (Medtronic, Inc., Minneapolis) that were filled with cpIL4-PE. Infusion within each catheter began within 24 h after catheter insertion at a very slow pace over a 4- to 8-day period (0.3–0.6 ml/h).

Preclinical studies guided the starting dose of cpIL4-PE. The first dose level of 0.2 µg/ml was 500 times lower than the maximum-tolerated dose (MTD; 100 µg/ml) that did not produce histological damage to the normal brain when administered in brain parenchyma in rats and 1/30 of the MTD injected intrathecally in monkeys. The dose was escalated one log to the next cohort of three patients and then a half log to the highest dose (6 µg/ml). The volume of fluid infused was determined on the basis of the tumor volume determined by MRI, including 1–2 cm margins of normal brain tissue. No symptoms of increased ICP (intracranial pressure) were observed during the infusion. However, cerebral edema causing corresponding signs and symptoms of ICP developed on days 27, 14, and 10 in three patients respectively. Four other patients developed increased ICP between 43 and 97 days post-cpIL4-PE infusion. Seven of nine patients experiencing increased ICP underwent craniotomy, which ameliorated these symptoms. Since three patients showed increased ICP within 30 days of infusion, it is possible that the amount of fluid administered contributed to this adverse event. Alternatively, cpIL4-PE might have been toxic to the normal brain. However, biopsy samples surrounding tumor resection did not show toxicity to the normal brain.[100] No apparent systemic toxicity based on serum chemistry and hematological studies occurred in any patient. Six of nine patients receiving infusions of cpIL4-PE showed tumor necrosis. Of six patients, one patient remained disease free for >18 months after the procedure (Figure 2.3). Based on these initial clinical results, we concluded that direct glioma injection of

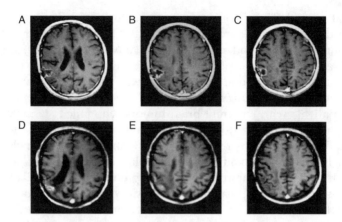

Figure 2.3   Complete regression of glioblastoma by cpIL4-PE (adapted from Ref. 100). Contrast brain MRI studies of a patient: three consecutive sections (A to C) show a cavity lesion in the right partial resection of a glioblastoma. Abnormal enhancement is present in the periphery of the lesion representing residual or recurrent tumor. There is no significant mass effect or edema. In D to F, images of comparable tomographic sections 9 months posttreatment show that the cavity has decreased in size. Residual enhancement remains in the regions of the tumor

cpIL4-PE was relatively safe without systemic toxicity, and caused necrosis of malignant gliomas that were refractory to conventional therapy.

Additional clinical trials were subsequently initiated at multiple centers in the United States and Germany to determine the MTD, volume, and safety of this agent when injected stereotactically.[102] This trial was designed as a dose-escalation trial of i.t. administration of cpIL4-PE in 31 patients with recurrent malignant glioma. Patients diagnosed with supratentorial grades 3 and 4 malignant glioma were verified histologically. Twenty-five patients were diagnosed with GBM, while six were diagnosed with anaplastic astrocytoma (AA) and assigned to one of four dose groups: 6 µg/ml × 40 ml, 9 µg/ml × 40 ml, 15 µg/ml × 40 ml, or 9 µg/ml × 100 ml of cpIL4-PE administered i.t. via stereotactically placed catheters. As in the previous pilot trial, all patients had adequate baseline organ function. Patients with subependymal or CSF disease, with anaplastic oligodendroglioma, with tumors involving the brainstem, cerebellum, or both hemispheres, with an active infection requiring treatment or with an unexplained febrile illness, who had received any form of radiation therapy or chemotherapy within 4 weeks of enrollment, or with systemic diseases that may have been associated with unacceptable anesthetic or operative risk were excluded. Similar to the previous trial, no systemic toxicity was apparent in any patients. Adverse effects noted were primarily limited to the central nervous system and appeared to be related to increased cerebral edema seen after drug administration. In all cases, the edema was treatable by medical or surgical means and neurological deficits were for the most part transient in nature. There were no deaths attributable to cpIL4-PE. Drug-related grades 3 or 4 CNS toxicity according to NCI Common Toxicity Criteria, was seen in 39% of patients in all dose groups. Serum levels of cpIL4-PE assessed by ELISA (sensitivity $\geq$ 5 ng/ml) were undetectable suggesting that intact cpIL4-PE did not enter the systemic circulation in appreciable quantities, however IgG antibody to the PE domain increased. Because no cpIL4-PE could be detected in the plasma, it is unclear whether the detected antibody to PE was against intact cpIL4-PE or fragments of cpIL4-PE.

The evidence of tumor necrosis was evaluated by changes in gadolinium enhancement by MRI. Post-infusion contrast MRI scans showed a distinct region of decreased signal intensity consistent with possible tumor necrosis and decreased contrast enhancement immediately after the end of infusion. Although tumor necrosis was not confirmed by biological examination, in our previous study, cpIL4-PE induced changes in gadolinium enhancement representing positive tumor necrosis was confirmed by histological examination of tissues in several patients.[100] The MTD of cpIL-4PE has yet to be defined in this study. In addition, because of known difficulties in the assessment of biological responses in the recurrent glioma setting no clear-cut drug effect was identified. To resolve these issues, a new protocol was designed, which stipulated CED of cpIL-4PE followed by tumor resection approximately 3 weeks later so MRI results may be confirmed by visual and histological examination. This protocol has been initiated. Our studies underscore the difficulty in designing a trial for recurrent glioma therapy and suggest that determining the optimum delivery of an agent to tumors is as critical as identification of a novel target, such as IL-4R.

Concluding remarks

In this chapter, we have summarized our discovery of the overexpression of IL-4R in human solid tumor cell lines including malignant glioma cells. From 76% to 100% of malignant astrocytoma samples obtained from patients undergoing surgical resection, and all glioblastoma cell lines and primary glioma cultures examined, were positive for the expression of IL-4R as determined by RT-PCR, Southern blot, and immunohistochemical analyses. In contrast, 6 of 11 normal brain tissues were found to be positive for IL-4R mRNA, however, protein expression could not be documented on normal brain sections by immunohistochemistry. We also determined that the structure of IL-4R on glioma cells is different from normal immune cells.

To target brain tumor cells, we have produced three generations of IL-4R-targeted cytotoxins. Among these proteins, a circular permuted IL-4 cytotoxin IL4(38-37)-PE38KDEL was more effective against brain tumors compared to first-generation IL-4 cytotoxin IL4-PE[4E]. The circular permutation of IL-4 not only improved the binding affinity of IL-4 cytotoxin but also improved cytotoxicity to tumor cells. This is a desirable characteristic of a chimeric molecule that is expected to result in a more effective targeting agent with no increase in nonspecific toxicity.

As cpIL4-PE was highly cytotoxic to IL-4R bearing brain cancer cell lines and primary cell cultures *in vitro*, we investigated whether it is also active *in vivo* in animal models of glioma tumors. Our extensive experiments successfully demonstrated that administration of cpIL4-PE to human tumor bearing mice can cause significant antitumor activity without any evidence of visible toxicity. Intratumoral injection of cpIL4-PE was most effective in eliminating gliomas established at subcutaneous sites in the flanks of nude mice. This is because intratumor administration of drug resulted in the saturation of the tumor bed, with subsequent complete eradication of established large tumors. Thus, it is reasonable to believe that intraglioma administration of cpIL4-PE will saturate the entire tumor bed and, because of bulk flow action, this drug may also be able to saturate the immediate proximity of the tumor bed, which contains tumor cells that cause recurrence.

Although we did not discuss here in detail, cancer cells that express low levels of IL-4R are less sensitive to the cytotoxic effect of cpIL4-PE.[27] Our recent data have shown that gene transfer of IL-4Rα chain in tumor cells can highly sensitize these tumors to the cytotoxic effect of cpIL4-PE both *in vitro* and *in vivo*.[91,103,104] Using this approach, it is possible that localized tumor can be treated with a minimal dose of cpIL4-PE without nonspecific toxicity. Additional studies are currently being undertaken in our laboratory to take advantage of this knowledge to develop additional therapies for human glioma.

Finally, two Phase I clinical trials for the treatment of recurrent grade IV astrocytoma by intratumoral administration of cpIL4-PE have been completed. Preliminary clinical results suggest that cpIL4-PE can be safely administered without systemic toxicities. The treatment-related adverse events were limited to the central nervous system. The MTD of 6 μg/ml × 40 ml of cpIL4-PE was determined. Evidence of tumor response was shown in the decreased signal density by MRI consistent with tumor necrosis following treatment. As intratumoral administration bypassed systemic toxicity and because glioblastoma multiforme is not generally a systemic disease, intratumoral injection of cpIL4-PE offers a promising new agent for brain tumor therapy.

Acknowledgments

Studies reported in this chapter were performed in collaboration with Ms. Pamela Dover, Drs. S. Rafat Husain, Nicholas I. Obiri, Bharat Joshi, and Takashi Murata of the Laboratory of Molecular Tumor Biology, Center for Biologics Evaluation and Research and Dr. Ira Pastan of the Laboratory of Molecular Biology, National Cancer Institute. The first clinical trial was performed in collaboration with Dr. Robert Rand of the John Wayne Cancer Institute, Santa Monica, California. Additional clinical trials were conducted at various centers in the Unites States and Germany. Authors would like to acknowledge the important contributions of the following individuals involved in various preclnical and clinical studies: Drs. Friedrich Weber,

Anthony Asher, Richard Bucholz, Mitchel Berger, Michael Prados, Jeff Bruce, Walter Hall, Nikolas Rainov, Manford Westphal, Ronald Warnick, Roger Williams, Vijay Hingorani, Susan Chang, Richard Maki, Robert Rand, and Hans deHan. These studies were conducted as part of the collaboration between FDA and Neurocrine Bioscience Inc. under a Cooperative Research and Development Agreement (CRADA). The views presented in this article do not necessarily reflect those of the Food and Drug Administration.

References

1. W.E. Paul, "Interleukin-4: A prototypic immunoregulatory lymphokine," *Blood* 77, 1990, 1859–1870.
2. R.K. Puri and J.P. Siegel, "Interleukin-4 and cancer therapy," *Cancer Invest.* 11, 1993, 473–486.
3. R.K. Puri, "Structure and function of interleukin 4 and its receptors," in R. Kurzrock and M. Talpaz (eds.), *Cytokines: interleukins and their receptors*, Massachusetts: Kluwer Academic Publishers, 1995, pp. 43–186.
4. M. Toi, R. Bicknel, and A.L. Harris, "Inhibition of colon and breast carcinoma cell growth by interleukin-4," *Cancer Res.* 52, 1992, 275–279.
5. R.I. Tepper, P.K. Pattengale, and P. Leder, "Murine interleukin-4 displays potent anti-tumor activity *in vivo*," *Cell* 57, 1989, 503–512.
6. R.I. Tepper, R.L. Coffman, and P. Leder, "An eosinophil-dependent mechanism for the antitumor effect of interleukin-4," *Science* 257, 1992, 548–551.
7. T. Morisaki, D.H. Yuzuki, R.T. Lin, L.J. Foshag, D.L. Morton, and D.S.B. Hoon, "Interleukin-4 receptor expression and groeth inhibition of gastric carcinoma by interleukin-4," *Cancer Res.* 52, 1992, 6059–6065.
8. M.S. Topp, M. Koenigsmann, A. Mire-Sluis, D. Oberberg, F. Eitelbach, Z. Von Marschall, M. Notter, B. Reufi, H. Stein, E. Thiel, and W.E. Berdel, "Recombinant human interleukin-4 inhibits growth of some human lung tumor cell lines *in vitro* and *in vivo*," *Blood* 82, 1993, 2837–2844.
9. O.V. Volpert, T. Fong, A.E. Koch, J.D. Peterson, C. Waltenbaugh, R.I. Tepper, and N.P. Bouck, "Inhibition of angiogenesis by interleukin 4," *J. Exp. Med.* 188, 1998, 1039–1046.
10. M.B. Atkins, G. Vachino, H. Tilg, H.J. Karp, N.J. Robert, K. Kappler, and J.W. Mier, "Phase I evaluation of thrice daily intravenous bolus interleukin-4 in patients with refractory malignancy," *J. Clin. Oncol.* 10, 1992, 1802–1809.
11. M.H. Gilleece, J.H. Scarffe, A. Ghosh, C.M. Heyworth, E. Bonnem, N. Tesla, P. Stern, and T.M. Dexter, "Recombinant human interleukin 4 (IL-4) given as daily subcutaneous injections — a Phase I dose escalation toxicity trial," *Br. J. Cancer* 66, 1992, 204–210.
12. J. Prendiville, N. Thatcher, M. Lind, R. Macintosh, A. Ghosh, P. Stern, and D. Crowther, "Recombinant human interleukin 4 (IL-4) administered by the intravenous and subcutaneous injections — a phase I toxicity study and pharmacokinetics analysis," *Eur. J. Cancer* 29A, 1993, 1799–1807.
13. K. Kawakami, M. Kawakami, and R.K. Puri, "Overexpressed cell surface interleukin-4 receptor molecules can be successfully targeted for antitumor cytotoxin therapy," *Crit. Rev. Immunol.* 21, 2001, 299–310.
14. R.K. Puri, M. Ogata, P. Leland, G.M. Feldman, and I. Pastan, "Expression of high affinity IL4 receptors on murine sarcoma cells and receptor mediated

cytotoxicity of tumor cells to chimeric protein between IL-4 and *Pseudomonas* extoxin," *Cancer Res.* 51, 1991, 3011–3017.

15. N.I. Obiri, G. Hillman, G.P. Haas, S. Sud, and R.K. Puri, "Expression of high affinity interleukin-4 receptors on human renal cell carcinoma cells and inhibition of tumor cell growth *in vitro* by interleukin-4," *J. Clin. Invest.* 91, 1993, 88–93.

16. P. Valent, J. Besemer, K. Kishi, F. diPadova, K. Geissler, K. Lechner, and P. Bettelheim, "Human basophils express interleukin-4 receptors," *Blood* 76, 1990, 1734–1738.

17. L.S. Park, D. Friend, H.M. Sassenfeld, and D.L. Urdal, "Characterization of the human B cell stimulatory factor 1 receptor," *J. Exp. Med.* 166, 1987, 476–478.

18. J.W. Lowenthal, B.E. Castle, J. Christiansen, J. Schreurs, D. Rennick, N. Arai, P. Hoy, Y. Takabe, and M. Howard, "Expression of high affinity receptors for murine interleukin 4 (BSF-1) on hematopoietic and nonhematopoietic cells," *J. Immunol.* 140, 1988, 456–464.

19. K. Nakajima, T. Hirano, K. Koyama, and T. Kishimoto, "Detection of receptors for B cell stimulatory factor 1 (BSF 1): Presence of functional receptors on CBA/N splenic B cells," *J. Immunol.* 139, 1987, 774–779.

20. C.F. Zuber, J.-P. Galizzi, A. Valle, N. Harada, M. Howard, and J. Banchereau, "Interleukin 4 receptors on normal human B lymphocytes: characterization and regulation," *Eur. J. Immunol.* 20, 1990, 551–555.

21. C.F. Zuber, J.-P. Galizzi, A. Valle, N. Harada, I. Durand, and J. Banchereau, "Interleukin-4 Receptors on human blood mononuclear cells," *Cell. Immunol.* 129, 1990, 329–340.

22. S.R. Husain, P. Gill, R.J. Kreitman, I. Pastan, and R.K. Puri, "Interleukin-4 receptor expression on AIDS-associated Kaposi's sarcoma cells and their targeting by a chimeric protein comprised of circularly permuted interleukin-4 and *Pseudomonas* exotoxin," *Mol. Med.* 3, 1997, 327–338.

23. N.I. Obiri and R.K. Puri, "Characterization of interleukin-4 receptors expressed on human renal cell carcinoma cells," *Oncol. Res.* 6, 1994, 419–427.

24. N.I. Obiri, J.P. Siegel, F. Varricchio, and R.K. Puri, "Expression and function of high affinity interleukin-4 receptors on human melanoma, ovarian and breast carcinoma cells," *Clin. Exp. Immunol.* 95, 1994, 148–155.

25. F. Varricchio, S.R. Husain, P. Leland, P. Gill, and R.K. Puri, "Interleukin-4 receptor expression *in vivo* on human AIDS-related Kaposi's sarcoma," *Oncol. Res.* 9, 1997, 495–503.

26. P. Leland, J. Taguchi, S.R. Husain, R.J. Kreitman, I. Pastan, and R.K. Puri, "Human breast carcinoma cells express type II IL-4 receptors and are sensitive to antitumor activity of a chimeric IL-4-*Pseudomonas* exptoxin fusion protein *in vitro* and *in vivo*," *Mol. Med.* 6, 2000, 165–178.

27. R.K. Puri, P. Leland, R.J. Kreitman, and I. Pastan, "Human neurological cancer cells express interleukin-4 (IL-4) receptors which are targets for the toxic effects of IL4-Pseudomonas exotoxin chimeric protein," *Int. J. Cancer* 58, 1994, 574–581.

28. R.K. Puri, D.S. Hoon, P. Leland, P. Snoy, R.W. Rand, I. Pastan, and R.J. Kreitman, "Preclinical development of a recombinant toxin containing circularly permuted interleukin-4 and truncated *Pseudomonas* exotoxin for therapy of malignant astrocytoma," *Cancer Res.* 56, 1996, 5631–5637.

29. R. Mehrotra, F. Varricchio, S.R. Husain, and R.K. Puri, "Head and neck cancers, but not benign lesions, express interleukin-4 receptors *in situ*," *Oncol. Rep.* 5, 1998, 45–48.

30. K. Kawakami, P. Leland, and R.K. Puri, "Structure, function, and targeting of interleukin 4 receptors on human head and neck cancer cells," *Cancer Res.* 60, 2000, 2981–2987.

31. K. Kawakami, M. Kawakami, S.R. Husain, and R.K. Puri, "Targeting interleukin-4 receptors for effective pancreatic cancer therapy," *Cancer Res.* 62, 2002, 3575–3580.

32. M. Kawakami, K. Kawakami, V.A. Stepensky, R.A. Maki, H. Robin, W. Muller, S.R. Husain, and R.K. Puri, "Interleukin 4 receptor on human lung cancer: a molecular target for cytotoxin therapy," *Clin. Cancer Res.* 8, 2002, 3503–3511.

33. S.R. Husain, K. Kawakami, M. Kawakami, and R.K. Puri, "Interleukin-4 receptor-targeted cytotoxin therapy of androgen-dependent and -independent prostate carcinoma in xenograft models," *Mol. Cancer Ther.* 2, 2003, 245–254.

34. B.H. Joshi, G.E. Plautz, and R.K. Puri, "Interleukin-13 receptor α chain: a novel tumor-associated transmembrane protein in primary explants of human malignant gliomas," *Cancer Res.* 60, 2000, 1168–1172.

35. B.H. Joshi, P. Leland, A. Asher, R.A. Prayson, F. Varricchio, and R.K. Puri, "*In situ* expression of interleukin-4 (IL-4) receptors in human brain tumors and cytotoxicity of a recombinant IL-4 cytotoxin in primary glioblastoma cell cultures," *Cancer Res.* 61, 2001, 8058–8061.

36. R.L. Idzerda, C.J. March, B. Mosley, S.D. Lyman, T.V., Bos, S.D. Gimpel, W.S. Din, K.H. Grabstein, M.B. Widmer, L.S. Park, D. Cosman, and M.P. Beckmann, "Human interleukin 4 receptor confers biological responsiveness and defines a novel receptor superfamily," *J. Exp. Med.* 171, 1990, 861–873.

37. S. Eyckerman, A. Verhee, J.V. derHeyden, I. Lemmens, X.V. Ostade, J. Vandekerckhove, and J. Tavernier, "Design and application of a cytokine-receptor-based interaction trap," *Nat. Cell Biol.* 3, 2001, 1114–1119.

38. S.M. Russell, A.D. Keegan, N. Harada, Y. Nakamura, M. Noguchi, P. Leland, M.C. Friedmann, A. Miyajima, R.K. Puri, W.E. Paul, and W.J. Leonard, "Interleukin-2 receptor gamma chain: a functional component of the interleukin-4 receptor," *Science* 262, 1993, 1880–1883.

39. M. Kondo, T. Takeshita, N. Ishii, M. Nakamura, S. Watanabe, K. Arai, and K. Sugamura, "Sharing of interleukin-2 (IL-2) receptor γc chain between receptors for IL-2 and IL-4," *Science* 262, 1993, 1874–1877.

40. M. Noguchi, Y. Nakamura, S.M. Russell, S.F. Ziegler, M. Tsang, X. Cao, and W.J. Leonard, "Interleukin-2 receptor γ chain: a functional component of interleukin-7 receptor," *Science* 262, 1993, 1877–1880.

41. S.M. Russell, M. Johnston, M. Noguchi, M. Kawamura, C. Bacon, M. Friedman, M. Berg, B. McVicar, B. Witthuhn, and O. Silvennoinen, "Interaction of IL-2 receptor beta and gamma chain with JAK 1 and JAK 3: implications for XSCID and XCID," *Science* 266, 1994, 1042–1045.

42. J.G. Giri, M. Ahdieh, J. Eisenman, K. Snanebeck, K. Grabstein, S. Kumaki, A. Naman, L.S. Park, D. Cosman, and D. Anderson, "Utilization of beta and gamma chains of the IL-2 receptor by novel cytokine IL-15," *EMBO J.* 13, 1994, 2822–2830.

43. H. Asao, C. Okuyama, S. Kumaki, N. Ishii, S. Tsuchiya, D. Foster, and K. Sugamura, "Cutting edge: the common gamma-chain is an indispensable subunit of the IL-21 receptor complex," *J. Immunol.* 167, 2001, 1–5.

44. N.I. Obiri, W. Debinski, W.J. Leonard, and R.K. Puri, "Receptor for interleukin 13: interaction with interleukin 4 by a mechanism that does not involve the common γ chain shared by receptors for interleukins 2, 4, 7, 9, and 15," *J. Biol. Chem.* 270, 1995, 8797–8804.

45. T. Murata, P.D. Noguchi, and R.K. Puri, "Receptors for interleukin (IL)-4 do not associate with the common γ chain, and IL-4 induce the phosphorylation of JAK2 tyrosine kinase in human colon carcinoma cells," *J. Biol. Chem.* 270, 1995, 30829–30836.

46. M.J. Aman, N. Tayebi, N.I. Obiri, R.K. Puri, W.S. Modi, and W.J. Leonard, "cDNA cloning and characterization of the human interleukin-13 receptor α chain," *J. Biol. Chem.* 271, 1996, 29265–29270.

47. D.J. Hilton, J.-G. Zhang, D. Metcalf, W.S. Alexander, N.A. Nicola, and T.A. Willson, "Cloning and characterization of a binding subunit of the interleukin 13 receptor that is also a component of the interleukin 4 receptor," *Proc. Natl. Acad. Sci. USA* 93, 1996, 497–501.

48. M. Kawakami, P. Leland, K. Kawakami, and R.K. Puri, "Mutation and functional analysis of IL-13 receptors in human malignant glioma cells," *Oncol. Res.* 12, 2001, 459–467.

49. S.M. Zurawski, P. Chomarat, O. Djossou, C. Bidaud, A.N.J. McKenzie, P. Miossec, J. Banchereau, and G. Zurawski, "The primary binding subunit of the human interleukin-4 receptor is also a component of the interleukin-13 receptor," *J. Biol. Chem.* 270, 1995, 13869–13878.

50. N.I. Obiri, P. Leland, T. Murata, W. Debinski, and R.K. Puri, "The interleukin-13 receptor structure differs on various cell types and may share more than one component with interleukin-4 receptor," *J. Immunol.* 158, 1997, 756–764.

51. T. Murata, N.I. Obiri, and R.K. Puri, "Structure of and signal transduction through interleukin 4 and interleukin 13 receptors," *Int. J. Mol. Med.* 1, 1998, 551–557.

52. T. Murata, N.I. Obiri, and R.K. Puri, "Human ovarian carcinoma cell lines express IL-4 and IL-13 receptors: comparison between IL-4 and IL-13 signaling," *Int. J. Cancer* 70, 1997, 230–240.

53. A.D. Keegan, J.A. Johnston, P.J. Tortolani, L.J. McReynolds, C. Kinzer, J.J. O'Shea, and W.E. Paul, "Similarities and differences in signal transduction by interleukin 4 and interleukin 13: analysis of Janus kinase activation," *Proc. Natl. Acad. Sci. USA* 92, 1995, 7681–7685.

54. M.G. Malabarba, H. Rui, H.H.J. Deutsch, J. Chung, F.S. Kalthoff, W.L. Farrar, and R.A. Kirken, "Interleukin-13 is a potent activator of JAK3 and STAT6 in cells expressing interleukin-2 receptor-γ and interleukin-4 receptor-α," *Biochem. J.* 319, 1996, 865–872.

55. D.C. Webb, A.N.J. McKenzie, A.M.L. Koskinen, M. Yang, J. Mattes, and P.S. Foster, "Integrated signals between IL-13, IL-4, and IL-5 regulate airways hyperreactivity," *J. Immunol.* 165, 2000, 108–113.

56. K. Nelms, A.D. Keegan, J. Zamorano, J.J. Ryan, and W.E. Paul, "The IL-4 receptor: signaling mechanisms and biologic functions," *Annu. Rev. Immunol.* 17, 1999, 701–738.

57. T. Murata, J. Taguchi, and R.K. Puri, "Interleukin-13 receptor α' chain but not α chain: a functional component of interleukin-4 receptor," *Blood* 91, 1998, 3884–3891.

58. T. Murata and R.K. Puri, "Comparison of IL-13 and IL-4 induced signaling in EBV immortalized human B cells," *Cell. Immunol.* 175, 1997, 33–40.

59. N.I. Obiri, T. Murata, W. Debinski, and R.K. Puri, "Modulation of interleukin-13 binding and signaling in human renal cell carcinoma cells by the γc chain of the IL-2 receptor," *J. Biol. Chem.* 272, 1997, 20251–20258.

60. J.N. Myers, S. Yasumura, Y. Suminami, H. Hirabayashi, W.-C. Lin, J.T. Johnson, M.T. Lotze, and T.L. Whiteside, "Growth stimulation of human head and neck squamous cells carcinoma cell lines by interleukin 4," *Clin. Cancer Res.* 2, 1996, 127–135.

61. R. Essner, Y. Huynh, T. Nguyen, D. Morton, and D.S. Hoon, "Functional interleukin 4 receptor and interleukin 2 receptor common gamma-chain on human hon-small cell lung cancers: novel targets for immune therapy," *J. Thorac. Cardiovasc. Surg.* 119, 2000, 10–20.

62. N.I. Obiri, N. Tandon, and R.K. Puri, "Upregulation of intracellular adhesion molecule-1 on human renal cell carcinoma cells by interleukin-4," *Int. J. Cancer* 61, 1995, 635–642.

63. P. Leland, N.I. Obiri, B.B. Aggarwal, and R.K. Puri, "Suramin blocks binding of interleukin-4 to its receptors and interleukin-4-induced mitogenic response," *Oncol. Res.* 7, 1995, 227–235.

64. S.R. Husain, P. Leland, B.B. Aggarwal, and R.K. Puri, "Transcriptional upregulation of interleukin 4 receptors by human immunodeficiency virus type 1 *tat* gene," *AIDS Res. Human Retroviruses* 12, 1996, 1349–1356.

65. D. Caput, P. Laurent, M. Kaghad, J.M. Lelias, S. Lefort, N. Vita, and P. Ferrara, "Cloning and characterization of a specific interleukin (IL)-13 binding protein structurally related to the IL-5 receptor α chain," *J. Biol. Chem.* 271, 1996, 16921–16926.

66. D.D. Donaldson, M.J. Whitters, L.J. Fitz, T.Y. Neben, H. Finnerty, S.L. Henderson, R.M. O'Hara Jr., D.R. Beier, K.J. Turner, C.R. Wood, and M. Collins, "The murine IL-13 receptor α2: molecular cloning, characterization, and comparison with murine IL-13 receptor α1," *J. Immunol.* 161, 1998, 2317–2324.

67. K. Kawakami, J. Taguchi, T. Murata, and R.K. Puri, "The interleukin-13 receptor α2 chain: an essential component for binding and internalization but not for IL-13 induced signal transduction through the STAT6 pathway," *Blood* 97, 2001, 2673–2679.

68. K. Kawakami, F. Takeshita, and R.K. Puri, "Identification of distinct roles for a dileucine and a tyrosine internalization motif in the interleukin (IL)-13 binding component IL-13 receptor α2 chain," *J. Biol. Chem.* 276, 2001, 25114–25120.

69. N. Feng, S.M. Lugli, B. Schnyder, J.-F.M. Gauchat, P. Graber, E. Schlagenhauf, B. Schnarr, M. Wiederkehr-Adam, A. Duschl, M.H. Heim, R.A. Lutz, and R. Moser, "The interleukin-4/interleukin-13 receptor of human synovial fibroblasts: overexpression of the nonsignaling interleukin-13 receptor α2," *Lab. Invest.* 78, 1998, 591–602.

70. J. Bernard, D. Treton, C. Vermot-Desroches, C. Boden, P. Horellou, E. Angevin, P. Galanaud, J. Wijdenes, and Y. Richard, "Expression of interleukin 13 receptor in glioma and renal cell carcinoma: IL-13Ralpha2 as a decoy receptor for IL-13," *Lab. Invest.* 81, 2001, 1223–1231.

71. K. Kawakami, M. Kawakami, P.J. Snoy, S.R. Husain, and R.K. Puri, "*In vivo* overexpression of IL-13 receptor α2 chain inhibits tumorigenicity of human breast and pancreatic tumors in immunodeficient mice," *J. Exp. Med.* 194, 2001, 1743–1754.

72. S.O. Rahaman, P. Sharma, P.C. Harbor, M.J. Aman, M.A. Vogelbaum, and S.J. Haque, "IL-13Rα2, a decoy receptor for IL-13 acts as an inhibitor of IL-4-dependent signal transduction in glioblastoma cells," *Cancer Res.* 62, 2002, 1103–1109.

73. E.S. Vitetta, R.J. Fulton, R.D. May, M. Till, and J.M. Uhr, "Redesigning nature's poisons to create antitumor agents," *Science* 238, 1988, 1098–1101.

74. I. Pastan and D.J. FitzGerald, "Recombinant toxins for cancer treatment," *Science* 254, 1991, 1173–1177.

75. I. Pastan, V. Chaudhary, and D.J. FitzGerald, "Recombinant toxins as novel therapeutic agents," *Annu. Rev. Biochem.* 61, 1992, 331–354.

76. R.J. Kreitman, "Immunotoxins in cancer therapy," *Curr. Opin. Immunol.* 11, 1999, 570–578.

77. A.E. Frankel, R.J. Kreitman, and E.A. Sausville, "Targeted toxins," *Clin. Cancer Res.* 6, 2000, 326–334.

78. Y. Reiter, "Recombinant immunotoxins in targeted cancer cell therapy," *Adv. Cancer Res.* 81, 2001, 93–124.

79. R.K. Puri, "Development of a recombinant interleukin-4-*Pseudomonas* exotoxin for therapy of glioblastoma," *Toxicol. Pathol.* 27, 1999, 53–57.

80. W. Debinski, R.K. Puri, R.J. Kreitman, and I. Pastan, "A wide range of human cancers express interelukin-4 receptors that can be targeted with chimeric toxin composed of IL-4 and Pseudomonas exotoxin," *J. Biol. Chem.* 268, 1993, 14065–14070.

81. R.K. Puri, W. Debinski, N.I. Obiri, R.J. Kreitman, and I. Pastan, "Human renal cell carcinoma cells are sensitive to the cytotoxic effect of a chimeric protein comprised of human interleukin-4 and Pseudomonas exotoxin," *Cell. Immunol.* 154, 1994, 369–379.

82. R.J. Kreitman, R.K. Puri, and I. Pastan, "A circularly permuted recombinant interleukin 4 toxin with increased activity," *Proc. Natl. Acad. Sci. USA* 91, 1994, 6889–6893.

83. R.J. Kreitman, R.K. Puri, and I. Pastan, "Increased antitumor activity of a circularly permuted interleukin 4-toxin in mice with interleukin 4 receptor-bearing human carcinoma," *Cancer Res.* 55, 1995, 3357–3363.

84. R.K. Puri, P. Leland, N.I. Obiri, S.R. Husain, J. Mule, I. Pastan, and R.J. Kreitman, "An improved circularly permuted interleukin 4-toxin is highly cytotoxic to human renal cell carcinoma cells," *Cell. Immunol.* 171, 1996, 80–86.

85. P. Hafkemeyer, U. Brinkmann, M.M. Gottesman, and I. Pastan, "Apoptosis induced by *Pseudomonas* exotoxin: a sensitive and rapid marker for gene delivery *in vivo*," *Hum. Gene Ther.* 10, 1999, 923–934.

86. A. Keppler-Hafkemeyer, R.J. Kreitman, and I. Pastan, "Apoptosis induced by immunotoxins used in the treatment of hematologic malignancies," *Int. J. Cancer* 87, 2000, 86–94.

87. M. Kawakami, K. Kawakami, and R.K. Puri, "Apoptotic pathways of cell death induced by an interleukin-13 receptor targeted recombinant cytotoxin in head and neck cancer cells," *Cancer Immunol. Immunother.* 50, 2002, 691–700.

88. M. Kawakami, K. Kawakami, and R.K. Puri, "Intratumoral administration of interleukin 13 receptor-targeted cytotoxin induces apoptotic cell death in human malignant glioma tumor xenografts," *Mol. Cancer Ther.* 1, 2002, 999–1007.

89. M. Kawakami, K. Kawakami, and R.K. Puri, "Tumor regression mechanisms by interleukin-13 receptor-targeted cancer therapy involve apoptotic pathways," *Int. J. Cancer* 103, 2003, 45–52.

90. S.R. Husain, N. Behari, R.J. Kreitman, and R.K. Puri, "Complete regression of established human glioblastoma tumor xenograft by interleukin-4 toxin therapy," *Cancer Res.* 58, 1998, 3649–3653.

91. K. Kawakami, M. Kawakami, P. Leland, and R.K. Puri, "Internalization property of interleukin-4 receptor α chain increases cytotoxic effect of interleukin-4 receptor-targeted cytotoxin in cancer cells," *Clin. Cancer Res.* 8, 2002, 258–266.

92. V.H. Freedman and S.I. Shin, "Cellular tumorigenicity in nude mice: correlation with cell growth in semi-solid medium," *Cell* 3, 1974, 355–359.

93. K. Kawakami, B.H. Joshi, and R.K. Puri, "Sensitization of cancer cells to interleukin 13-*Pseudomonas* exotoxin-induced cell death by gene transfer of interleukin 13 receptor α chain," *Hum. Gene Ther.* 11, 2000, 1829–1835.

94. S.R. Husain, R.J. Kreitman, I. Pastan, and R.K. Puri, "Interleukin-4 receptor-directed cytotoxin therapy of AIDS-associated Kaposi's sarcoma tumors in xenografted model," *Nat. Med.* 5, 1999, 817–822.

95. S.E. Strome, K. Kawakami, D. Alejandro, S. Voss, J.L. Kasperbauer, D. Salomao, L. Chen, R. Maki, and R.K. Puri, "Interleukin 4 receptor-directed cytotoxin therapy for human head and neck squamous cell carcinoma in animal models," *Clin. Cancer Res.* 8, 2002, 281–286.

96. K. Kawakami, S.R. Husain, M. Kawakami, and R.K. Puri, "Improved anti-tumor activity and safety of interleukin-13 receptor targeted cytotoxin by systemic continuous administration in head and neck cancer xenograft model," *Mol. Med.* 8, 2002, 487–492.

97. B. Morrison and P. Leder. "A receptor binding domain of mouse interleukin-4 defined by a solid phase binding assay and *in vitro* mutagenesis," *J. Biol. Chem.* 267, 1992, 11957–11963.

98. K.A. Gossett, T.A. Barbolt, J.B. Cornacoff, D.J. Zelinger, and J.H. Dean, "Clinical pathologic alterations associated with subcutaneous administration of recombinant interleukin-4 to cynomolgus monkeys," *Toxicol. Pathol.* 21, 1993, 46–53.

99. J.B. Cornacoff, K.A. Gossett, T.A. Barbolt, and J.H. Dean, "Preclinical evaluation of recombinant interleukin-4," *Toxicol. Lett.* 64/65, 1992, 299–310.

100. R.W. Rand, R.J. Kreitman, N. Patronas, F. Varricchio, I. Pastan, and R.K. Puri, "Intratumoral administration of recombinant circularly permuted Interleukin-4-*Pseudomonas* exotoxin in patients with high-grade glioma," *Clin. Cancer Res.* 6, 2000, 2157–2165.

101. D.W. Laske, R.J. Youle, and E.H. Oldfield, "Tumor regression with regional distribution of the targeted toxin TF-CRM107 in patients with malignant brain tumors," *Nat. Med.* 3, 1997, 1362–1368.

102. A.L. Asher, F. Weber, R. Bucholz, M. Berger, M. Prados, J. Bruce, W. Hall, N.G. Rainov, M. Westphal, R. Warnick, R.L. Willaims, V.N. Hingorani, and R.K. Puri, "Tumor response, safety and tolerability of IL4-toxin (NBI-3001) in patients with recurrent malignant gliomas," 51st Annual Meeting of Congress of Neurological Surgeons, San Diego, California, September 29–October 4, 2001, Abstract no. 763, p. 140.

103. K. Kawakami, M. Kawakami, and R.K. Puri, "Cytokine receptor as a sensitizer for targeted cancer therapy," *Anti-Cancer Drugs* 13, 2002, 693–699.

104. K. Kawakami, M. Kawakami, S.R. Husain, and R.K. Puri, "Effect of interleukin-4 cytotoxin on breast tumor growth after *in vivo* gene transfer of IL-4Rα chain," *Clin. Cancer Res.* 9, 1826–1836, 2003.

105. K. Kawakami, M. Kawakami, B.H. Joshi, and R.K. Puri, "Interleukin-13 receptor-targeted cancer therapy in an immunodeficient animal model of human head and neck cancer," *Cancer Res.* 61, 2001, 6194–6200.

chapter three

Interleukin-13 receptors and development of IL-13-Pseudomonas exotoxin for human cancer therapy

Bharat H. Joshi, Syed R. Husain, Pamela Leland, and Raj K. Puri

Contents

Despite significant advances in cancer therapy, more than one million Americans are diagnosed with cancer each year and about half of those patients

0-4152-6365-4/05/$0.00+$1.50
© 2005 by CRC Press

die from this disease. An increased incidence has been observed for several types of malignancies over the past two decades because of poor response to currently available therapies. Targeted therapeutic approaches based on the discovery of tumor antigens or receptors appear to provide novel highly specific agents for the treatment and management of cancer. To develop these molecules, we have identified a unique tumor target in the form of interleukin-13 receptors (IL-13R) that are overexpressed on the cell surface of different solid human tumors including renal cell carcinoma (RCC), brain tumors including glioblastoma multiforme (GBM), AIDS-associated Kaposi's sarcoma, some squamous cell carcinoma of head and neck (SCCHN), and ovarian cancer. To target these receptors, we developed a chimeric fusion protein composed of IL-13 and *Pseudomonas* exotoxin (IL-13 cytotoxin or IL-13-PE) that specifically binds to IL-13 receptor positive cells and kills them by inducing apoptosis and necrosis in a dose-dependent manner. In addition, IL-13 cytotoxin can mediate significant antitumor activity *in vivo* in a murine model of human cancers. Administration of this cytotoxin either by intravenous, intraperitoneal, or intratumor routes caused significant regression of established human RCC and glioma tumors with complete responses, which were durable for a long period of time. Preclinical studies for safety and toxicity in mice, rats, and monkeys have demonstrated that the animals tolerated IL-13 cytotoxin well with minimal toxicity to vital organs. In addition, intra rat brain parenchyma administration of IL-13 cytotoxin at doses up to 100 μg/ml was well tolerated without any evidence of gross or microscopic necrosis, but at a higher dose (500 μg/ml), the animals developed localized necrosis in their brain.

Based on these and other studies, four Phase I/II clinical trials were initiated in patients with GBM and advanced metastatic RCC. In GBM patients, the first clinical trial involves convection-enhanced delivery (CED) of IL-13 cytotoxin into recurrent malignant glioma. IL-13 cytotoxin administration through this route appears to be well tolerated with no signs of neurotoxicity. The second clinical trial involves infusion of IL-13 cytotoxin by CED following tumor resection. The initial stage of the second study assessed histologic effects of the drug administered prior to resection. In the third study, IL-13 cytotoxin is infused by CED followed by tumor resection. For the RCC trial, IL-13 cytotoxin was administered intravenously in patients with advanced and progressing disease. Our clinical studies have demonstrated that IL-13 cytotoxin could be safely administered intravenously at a dose of up to 2 μg/kg q.o.d. ×3 in the RCC trial and intratumorally up to 4 μg/ml in the GBM trial; these trials are currently ongoing. Future Phase I/II clinical trials will examine safety and activity of IL-13 cytotoxin in other cancers expressing IL-13 receptors and Phase II/III studies will examine the efficacy of IL-13 cytotoxin in human RCC and GBM patients.

Introduction

Human malignant tumors differ fundamentally from normal or non-cancerous cells by displaying an identifiable antigenic behavior and genetic instability.

It reflects a constant turnover of new antigens in tumors as they develop and progress. The pattern is highly specific, generally not seen in normal, nontransformed tissues, which preserve a stable phenotypic or antigenic profile. The uncontrolled tumor growth is definitely a common biological characteristic of all malignant tumors. In particular, malignant brain tumors, especially glioblastoma multiforme (GBM) and renal cell carcinoma (RCC), the focus of this chapter, have a poor prognosis because of their diffuse infiltrative nature and marked cytologic heterogeneity, respectively. More than 35,000 adult Americans are diagnosed with brain tumors each year; 15 to 35% of these intracranial tumors are diagnosed as GBM (Greig et al. 1990; Ries et al. 2000) and approximately 13,300 of these patients die from brain tumors each year (States 1994). Renal cell carcinoma (RCC), is the third most common tumor of the urogenital tract. Its incidence has been increasing by 2 to 4% per year since the 1970s (Zucchi et al. 2003). Out of the 31,900 patients diagnosed with RCC in the United States each year, approximately 11,900 patients die of their disease (Jemal et al. 2003).

Despite the use of conventional therapies such as radiotherapy, immunotherapy, chemotherapy, or surgery, the median survival time for WHO-classified grade IV malignant glioma remains a dismal 47 weeks (Galanis et al. 1998). Approximately 30% of RCC patients with metastatic disease respond to conventional therapeutic regimens. Though surgical removal of the localized or metastatic tumors is considered to be an effective therapy resulting in long-term survival in some individuals, these malignancies in RCC patients are usually uncontrollable once it has spread beyond the kidney (Indolfi et al. 2003). New therapeutic approaches including biological therapy with interleukin-2 or interferon have resulted in some responses in a minority of patients, with occasional long-term survivors (Fossa et al. 1991; Fyfe et al. 1995; Guirguis et al. 2002; Hurst et al. 1999; Rosenberg 1997; Wagner et al. 1999). On the other hand, recent advances in translational research and development of new therapeutic approaches, such as gene therapy (Deen et al. 1993; Valery et al. 2002) and immunotherapy, have shown new promises for treating patients with malignant gliomas and renal cell carcinoma. These approaches could also be used in combination with standard therapies or independently. An increasing number of studies are currently exploring the utility of immunotherapy including manipulating the host's cellular and humoral immune system for the management and therapy of RCC (Fossa et al. 1991; Fyfe et al. 1995; Indolfi et al. 2003). Further, the use of anti-angiogenic factors or antibodies to these factors have also shown some promising results (Campbell and Marcus 2003; Dabrowska et al. 2001; Liu et al. 2003; Matsuoka et al. 2003; Oosterwijk et al. 2003; Scappaticci 2002).

Another approach for the therapy of human cancer has evolved over the past two decades. In this approach, tumor-specific antigens or receptors are targeted by conjugated or fused immunotoxins or cytotoxins, which comprise an antibody, a fragment or a ligand, and a toxin moiety derived either from plant or bacteria or sometimes a chemical toxin (Frankel et al. 2000a; Frankel et al. 2000b; Frankel et al. 2002; Goldberg et al. 1995; Grossbard et al. 1993; Husain et al. 1998; Joshi et al. 2000; Kawakami et al. 2003; Kreitman et al. 1999; Pai et al. 1996; Pai-Scherf et al. 1999; Pastan et al. 1995; Puri, 2000; Waldmann et al. 1992.). This approach has resulted in the licensure of two targeted agents for the therapy of cutaneous T-cell lymphoma and acute myelogenous leukemia (Hamann et al. 2002; van Der Velden et al. 2001). A similar approach may also be very useful for other cancers including malignant brain tumors and renal cell carcinoma.

In this chapter, we focus on preclinical and clinical studies in the development of a molecule, IL-13 cytotoxin (IL-13-PE38QQR) for targeting of human cancers. Most preclinical and clinical studies will focus on malignant gliomas and human RCC as these studies have advanced considerably. Testing of IL-13 cytotoxin was made possible by our initial discovery of the overexpression of Interleukin-13 (IL-13) receptors on human RCC cell lines (Obiri et al. 1995). We also describe the receptor structure and biological significance of these receptors in RCC.

IL-13 and its receptors

IL-13 is an approximately 12 kDa pleotropic cytokine originally described as a T-cell-derived growth factor and was cloned from activated T-cells (Brown et al. 1989; Minty et al. 1993). The IL-13 gene has a close structural and functional relationship with the IL-4 gene, and IL-13 protein is found to have a 30% identity in the amino acid sequence to IL-4 protein (Aversa et al. 1993; Zurawski et al. 1993). The known biological functions of IL-13 have expanded considerably over the past few years. Some of the major biological activities of IL-13 include inhibition of production of inflammatory cytokines by monocytes (de Waal Malefyt et al. 1993; Minty et al. 1993) and induction of anti-CD40-dependent IgG/IgE class switch and IgG and IgM synthesis by B-cells (Cocks et al. 1993; Defrance et al. 1994; McKenzie et al. 1993; Minty et al. 1993; Punnonen et al. 1993). IL-13 shares many overlapping biological activities with IL-4. However in contrast to IL-4, IL-13 does not seem to have an effect on resting or activated T-cells (Punnonen et al. 1993; Zurawski and de Vries 1994). In recent years IL-13 has evolved as a key mediator of inflammatory diseases including allergies and pulmonary asthma (Blease et al. 2002; Blease et al. 2001; Danahay et al. 2002; Jakubzick et al. 2003a; Jakubzick et al. 2003b; Jakubzick et al. 2002; Kraft et al. 2001; Sela 1999; Tomkinson et al. 2001; Vogel 1998a; Vogel 1998b; Zimmermann et al. 2003).

IL-13 binds to its plasma membrane receptors to produce a biological response. Our laboratory was the first to identify the overexpression of IL-13 receptors (IL-13R) on human renal cell carcinoma cells (Obiri et al. 1995). Later on, we demonstrated that many human tumor cell lines and tissues originating from malignant glioma, AIDS-associated Kaposi's sarcoma (AIDS-KS), squamous cell carcinoma of head and neck (SCCHN) and medulloblastoma express large numbers of IL-13R (Debinski et al. 1996; Debinski et al. 1995b; Husain et al. 1997; Joshi et al. 2003; Joshi et al. 2002; Joshi et al. 2000; Kawakami et al. 2001d; Obiri et al. 1995). In contrast, human B-cells, monocytes,

T-cells, and endothelial cells express undetectable or low numbers of these receptors (Obiri et al. 1997b).

Because RCC tumor cells expressed high numbers of IL-13R, the binding characteristics and structure of the IL-13R was studied in RCC and other tumor cells. Based on the binding studies, we first proposed that ^{125}I-IL-13 cross links to a single 60 to 70 kDa protein on RCC tumor cells (Obiri et al. 1995). We further proposed that the IL-13R comprises two proteins each of 65 to 70 kDa (Obiri et al. 1995). Indeed, two different chains of IL-13 receptor were later cloned. One of the IL-13 receptor chains was first cloned in a murine system and later a human homologue was cloned (Aman et al. 1996; Hilton et al. 1996). This chain is termed IL-13Rα1 (also known as IL-13R or IL-13R α'). This chain forms a complex with the primary IL-4 binding protein (IL-4Rα chain, also known as IL-4Rβ) for signal transduction (Idzerda et al. 1990; Murata et al. 1998c). The second chain of the IL-13 receptor was cloned from a human RCC cell line and found to be a 65 kDa protein (Caput et al. 1996; Kawakami et al. 2000b). This chain was found to have 50% sequence homology at the DNA level with IL-5Rα chain. This chain is termed IL-13Rα2 (also known as IL-13Rα). Interestingly, IL-13Rα1 and IL-13Rα2 chains have no sequence homology with each other. These studies confirmed our previous predictions that the IL-13R was composed of two chains of similar size (Obiri et al. 1995).

Based on radio-labeled IL-13 binding, displacement, and cross-linking studies, we proposed that IL-13R structure varies in different cell types and the IL-13R shares two chains (IL-4Rα and IL-13Rα1) with the IL-4R system (Murata et al. 1997a; Obiri et al. 1997a). On the basis of these results, we proposed that IL-13R exists as three different types in different cells. In type I IL-13R, IL-13Rα1, IL-13Rα2, and IL-4Rα chains are present. IL-13 can bind to all three chains, while IL-4 binds to only two proteins (IL-4Rα and IL-13Rα1). Because of these characteristics [125]I- IL-13 binding was inhibited by an excess of IL-13 but not by IL-4. However, [125]I- IL-4 binding was displaced by both IL-4 and IL-13. This type of IL-13R has been shown to be expressed in RCC, brain tumor, AIDS-KS, certain SCCHN, and ovarian cancers (Debinski et al. 1995a; Debinski et al. 1995b; Husain et al. 1997; Joshi et al. 2002; Joshi et al. 2000; Joshi et al. 2003; Kawakami et al. 2001d; Murata et al. 1998b; Obiri et al. 1995). In type II IL-13R, the IL-13Rα2 chain is not present and IL-13 forms a complex with IL-13Rα1 and IL-4Rα chains. This type of IL-13 receptor is expressed in certain colon carcinoma, epidermal carcinoma, breast carcinoma, prostate carcinoma, and pancreatic cancer cells (Kawakami et al. 2001a; Kawakami et al. 2001c; Maini et al. 1997; Murata et al. 1997a; Murata et al. 1996; Murata et al. 1995). Because of the absence of the IL-13Rα2 chain, ^{125}I-IL-13 and ^{125}I-IL-4 binding are displaced by both IL-4 and IL-13 (Murata et al. 1996; Murata et al. 1998b; Obiri et al. 1997a). In type III IL-13R, an additional component from the IL-2R complex (γc chain) that may modulate IL-13 binding is also present. This type of IL-13R is expressed on B-cells, monocytes, and TF-1 cells (Obiri et al. 1997a; Vita et al. 1995; Zurawski et al. 1995). Although γc does not appear to bind IL-13, it affects IL-13 binding and signal transduction in certain types of cells (Kuznetsov and Puri 1999; Obiri et al. 1997b).

The structure of the IL-13R was confirmed by reconstitution studies (Kawakami et al. 2001d). When IL-13Rα1 alone was expressed in CHO-K1 cells, IL-13 bound with low affinity. However, when IL-13Rα1 was coexpressed with IL-4Rα, a high-affinity receptor-signaling complex was formed as reported by us and others (Aman et al. 1996; Hilton et al. 1996). This receptor complex is expressed widely on both lymphoid and nonlymphoid cells and can also be activated by IL-4, thus accounting for the functional overlap between IL-4 and IL-13 (Zurawski et al. 1993). The second IL-13 binding chain as described earlier, IL-13Rα2, binds IL-13 with high affinity and, in contrast to IL-13Rα1, is also found as a soluble receptor in mouse serum and urine (Donaldson et al. 1998; Kawakami et al. 2001d; Kawakami et al. 2001e; Zhang et al. 1997). Structural differences between the cytoplasmic domains of IL-13Rα1 and IL-13Rα2 chains are distinct. The cytoplasmic region of murine IL-13Rα2 does not possess an obvious signaling motif or janus kinases (JAK) and signal transducer and activator of transcription (STAT) binding sequence (Donaldson et al. 1998), which raises the possibility that it is a dominant negative inhibitor or decoy receptor, as originally described for the IL-1 receptor type II (Colotta et al. 1993). To understand the role of IL-13Rα2 in regulating the biological activity of IL-13, mice with targeted deletion of IL-13Rα2 were recently generated (Chiaramonte et al. 2003; Wood et al. 2003). Basal serum IgE levels were elevated in IL-13Rα2$^{-/-}$ mice despite the fact that serum IL-13 was absent and IFN-γ production increased compared to wild type mice. These studies indicated that one of the primary functions of IL-13Rα2 is to limit IL-13 effector function *in vivo*, and this property of the IL-13Rα2 chain can be exploited in diseases where IL-13 plays a major role (Chiaramonte et al. 2003; Chiaramonte et al. 2001; Chiaramonte et al. 1999a; Chiaramonte et al. 1999b; Grunig et al. 1998; Kawakami et al. 2001c; Morse et al. 2002; Terabe et al. 2000; Wills-Karp et al. 1998).

Signal transduction through IL-13 receptors

Not only do the IL-4R and IL-13R systems share two chains (IL-4α and IL-13α1) with each other, both IL-4 and IL-13 have been shown to signal through JAK-STAT pathways. Similar to IL-4, IL-13 can phosphorylate and activate JAK1 and Tyk2 tyrosine kinases in hemopoietic cells, while it phosphorylates JAK1 and JAK2 tyrosine kinases in non-hemopoietic cells such as colon carcinoma and fibroblast cells (Murata et al. 1998a; Murata et al. 1996). In contrast to IL-4, IL-13 could not phosphorylate and activate JAK3 tyrosine kinase in any cell type (Murata and Puri 1997). JAK kinase phosphorylation by IL-13 leads to phosphorylation of IL-4Rα and 170 kDa insulin receptor substrate II (IRS-II) protein in hemopoietic cells and non-hemopoietic cells (Murata et al. 1996; Smerz-Bertling and Duschl 1995; Welham et al. 1995). Despite the differences in signaling in hemopoietic and non-hemopoietic cells, IL-13 phosphorylated and activated STAT6 protein as did IL-4

(Kawakami et al. 2001d; Kohler et al. 1995; Kohler et al. 1994; Lin et al. 1995; Murata and Puri 1997). In addition, we and others have reported that the IL-13Rα2 chain not only inhibits signaling through STAT6 pathway for IL-13R but also for IL-4R (Murata et al. 1997b; Rahaman et al. 2002; Wood et al. 2003).

The specificity between IL-4 and IL-13 signaling perhaps resides in the proximal JAK signaling pathway. Therefore, the JAK-STAT signaling pathway may serve as a useful target for pharmacologic intervention to interfere with the effects of both IL-4 and IL-13 (Kohler et al. 1994; Lin et al. 1995; Murata et al. 1996).

Function of IL-13R in RCC cells

The mechanism involved in the inhibition of IL-13 binding to its receptors and signaling by the γ_c chain in these cells was associated with the downregulation of constitutive expression of IL-13Rα2 and ICAM-1 in ML-RCC cells (Obiri et al. 1997b). A mathematical model that fits a set of kinetic and steady state data in control and receptor positive cells was selected from a set of possible models. This best-fit model predicted that (1) two different IL-13R are expressed on the cell membrane, (2) a minor fraction of IL-13R exist as micro clusters (homodimers and/or heterodimers without exogenous IL-13, (3) high morphological complexity of the γ_c negative control cell membrane affects the cooperativity phenomena of IL-13 binding, and (4) a large number of coreceptor molecules are present, which helps keep the ligand on the cell surface for a long period of time after fast IL-13 binding and provides a negative control for ligand binding via production of the high affinity inhibitor bound to IL-13. These results are important as they demonstrate that γ_c exerts dramatic changes in the kinetic mechanisms of IL-13 binding (Kuznetsov and Puri 1999).

To determine the functional significance of the expression of IL-13R on RCC cells, studies were performed to analyze the impact of IL-13 on the growth of RCC cells. IL-13 inhibited cellular proliferation of three human renal cell carcinoma cell lines in a concentration-dependent manner (Obiri et al. 1996). As IL-13 binds with IL-13Rα1 chain, it recruits IL-4Rα for signaling. Therefore, we determined whether blocking IL-4Rα by a specific antibody would inhibit the IL-13 effect. Our data demonstrated that anti-IL-4Rα antibody did not block the effect of IL-13 indicating that these effects were IL-4Rα independent (Obiri et al. 1995; Obiri et al. 1996). In addition, IL-13 also induced ICAM-1 expression on RCC cells indicating that IL-13R are functional (Obiri et al. 1997b). To further analyze the significance of IL-13R overexpression in tumor cells, sequencing and single nucleotide conformation polymorphism (SSCP) studies were performed (Kawakami et al. 2000b). Although these studies were performed in glioma cells, our studies demonstrated that IL-13R are not subject to gene rearrangement or mutations (Kawakami et al. 2000b). Nevertheless, the role of IL-13R in the tumorigenic process has not been ruled out.

Development of IL-13 Pseudomonas *exotoxin (IL-13-PE38QQR) to target IL-13 receptor positive tumor cells*

The discovery of overexpression of IL-13R on human renal cell carcinoma cells led us to develop the IL-13-PE38QQR fusion protein, which comprises IL-13 and a bacterial toxin, *Pseudomonas* exotoxin (PE). This PE toxin is a single chain bacterial protein with three major domains (Pastan et al. 1992). The N-terminally located domain Ia binds to the α2-macroglobulin receptor, and the ligand–receptor complex undergoes receptor-mediated endocytosis to allow intracellular routing and processing of the toxin. Domain II is a site of proteolytic cleavage that activates PE and is essential for catalyzing the translocation of the toxin into the cytosol. Domain III contains the REDLK sequence, which directs the processed fragment of PE to the endoplasmic reticulum and possesses an ADP ribosylation activity, which inhibits elongation factor 2 and results in cell death (Iglewski and Kabat 1975; Pastan et al. 1992; Pastan et al. 1995). Human IL-13 was linked with a mutated form of PE (PE38QQR), where three lysine residues at 590, 606, and 613 in truncated PE38 were substituted with two glutamine and one arginine respectively to construct IL-13-PE38QQR. It is also referred to as IL-13-PE or IL-13 cytotoxin. This protein was expressed in a prokaryotic (pET) expression system using ampicillin or kanamycin as a selection antibiotic (Figure 3.1) (Debinski et al. 1995a; Joshi et al. 2002). This recombinant protein was found to be highly cytotoxic (as determined by the inhibition of protein synthesis) to IL-13R positive RCC cells at a very low concentration (Puri et al. 1996). This molecule was found to be highly cytotoxic to other IL-13R positive human solid tumor cell lines derived from malignant glioma, AIDS-KS, SCCHN, and ovarian, prostate, and several human epithelial carcinomas such as colon and skin (Debinski et al. 1995a; Debinski et al. 1995b; Husain and Puri 2000; Kawakami et al. 2001a; Maini et al. 1997; Puri et al. 1996).

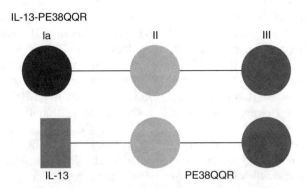

Figure 3.1   Construction of IL-13PE38QQR: N-terminally located domain Ia of truncated *Pseudomonas* exotoxin (PE38QQR) was replaced with human IL-13. Three lysine residues in the third domain of PE at position 590, 606, and 613 were substituted with two glutamine and one arginine respectively.

The level of cytotoxicity (IC_{50}, the concentration of the drug inhibiting protein synthesis by 50%) was dependent on the number of IL-13R on the cell surface. The higher the receptor number, the higher the cytotoxicity level (Puri et al. 1996). The cytotoxic activity of IL-13 cytotoxin was mediated through IL-13R as it was blocked when an excess of human IL-13 was included in the assay. The mechanism of tumor cell killing by IL-13 cytotoxin was investigated *in vitro* and *in vivo* in SCCHN and glioma tumors (Kawakami et al. 2002c; Kawakami et al. 2002d). IL-13 cytotoxin induced apoptosis at least in part in these tumors and two pathways (caspase and mitochondrial cytochrome C release) of apoptosis were operational (Kawakami et al. 2002d). Interestingly, IL-13 cytotoxin was more effective in mediating cytotoxicity in certain tumor types, which preferentially express high levels of IL-13Rα2 (Kawakami et al. 2001a; Kawakami et al. 2001d; Obiri et al. 1996).

Similar to RCC cell lines, IL-13 cytotoxin has been found to be highly cytotoxic to a variety of established glioma cell lines. The IL-13 toxin mediated specific cytotoxicity to glioma cell lines as they expressed up to 30,000 IL-13 receptors per cell (Debinski et al. 1995b; Husain and Puri 2003). The concentration of toxin at which 50% inhibition of protein synthesis was achieved (IC_{50}) in these cells generally ranged between <1 ng and 200 ng/ml (Debinski et al. 1995b; Husain and Puri 2003). In general, the number of IL-13 receptors positively correlated with the sensitivity of these cells to IL-13 cytotoxin. This study unequivocally demonstrated that the IL-13 cytotoxin is highly active on glioma cell lines and kills the tumor cells in a receptor number dependent manner. Consistent with the lack of expression of IL-13Rα2 chain, normal brain cells were not sensitive to IL-13 cytotoxin, suggesting that the IL-13Rα2 chain is responsible for IL-13 binding and internalization.

In developing targeted therapeutic protein-based approaches, the expression of a targeted receptor on normal tissues has always been a prime concern. Our RT-PCR results for normal human astrocytes (NHA) and normal brain tissues have shown that these cells expressed low levels of mRNA for the IL-13Rα2 chain. In addition, four other normal brain cell lines, derived from human oligodendrocytes, a neuronal cell line, and two cortex tissue-derived cell lines from a patient with encephalitis, also showed no or very low levels of IL-13Rα2 mRNA. Immunofluorescence analysis (IFA) analysis for IL-13Rα2 protein detection in NHA cells revealed that these cells express low levels of IL-13Rα2 protein on the cell surface confirming our RT-PCR results (Joshi et al. 2000). Thus, NHAs appeared to express type I IL-13R complex (comprised of IL-4Rα, 13Rα1, and IL-13Rα2). Consistent with the low-level expression of IL-13R on normal human brain tissues, IL-13 cytotoxin was not cytotoxic to NHAs. Similar to NHAs, normal endothelial and immune cells do not express or express very low levels of IL-13Rα2 (Husain et al. 1997; Obiri et al. 1995). Therefore, it is expected that IL-13 cytotoxin may not exert its deleterious effect on these cells. These studies have indicated that differential overexpression of IL-13R on tumor cells could serve as a primary target for IL-13 cytotoxin *in vivo*.

IL-13Rα2 chain is primarily responsible for IL-13 cytotoxin–mediated cytotoxicity

Our studies indicated that the IL-13Rα2 chain plays a major role in IL-13 cytotoxin–induced cytotoxicity of tumor cells. This hypothesis was confirmed by gene transfer studies. Plasmid-mediated gene transfer of IL-13Rα2 highly sensitized previously insensitive pancreatic, breast, head and neck, prostate, and glioblastoma cell lines to IL-13 cytotoxin *in vitro*. These *in vitro* results were confirmed by *in vivo* studies in a number of human tumor xenograft models (Kawakami et al. 2002a; Kawakami et al. 2002b; Kawakami et al. 2001b; Kawakami et al. 2000a). These *in vivo* studies revealed that the innate immune response was also activated mediating a robust tumor response. This is the first demonstration that an immunotoxin could induce cell killing in receptor positive tumors by recruiting the help of the innate immune response.

IL-13 receptors as targets for immunotherapy

To further study the therapeutic utility of *in vivo* IL-13Rα2 gene transfer in solid tumors followed by IL-13 cytotoxin therapy, the IL-13Rα2 gene was stably transfected into breast and pancreatic tumor cell lines. These cells were injected subcutaneously in immunodeficient animals (Kawakami et al. 2001c). Surprisingly, overexpression of the IL-13Rα2 chain in these tumor cells inhibited the growth of these tumors, while mock transfected control tumor cells formed tumors that increased in size (Kawakami et al. 2001c). IL-13Rα2 chain transfer also induced activation of innate immune response and production of anti-angiogenic factors at the tumor site. As IL-13 binds to IL-13Rα2 with high affinity, both IL-13 and IL-13R were hypothesized to be involved in tumorigenesis. A similar role of IL-13/IL-13R complex has also been reported by Terabe et al. (2000). Treatment of the host with an IL-13Rα2 extracellular domain-Fc fusion protein inhibited tumor regression in a tumor rejection-progression model indicating that IL-13 played a major role in tumor regrowth. As natural killer T (NKT) cells were found to produce IL-13, these cells seem to play a major role in tumor immunosurveillance. As IL-13 is a key Th2-derived cytokine, it is hypothesized that Th2 cells may be involved in evading immunosurveillance. This hypothesis was also proposed by several other studies. It was demonstrated that IL-4, IL-13, and STAT6 were responsible for mediating transplantable tumor rejection (Kacha et al. 2000; Ostrand-Rosenberg et al. 2000; Terabe et al. 2000). Mice with STAT6 knock out, which mediate terminal signaling through IL-4 and IL-13 receptors, rejected the transplanted tumor cells, while control animals formed enlarged tumors. Enhancement of tumor-specific IFN-γ production and CTL activity in the absence of STAT6 was proposed as the primary explanation for the potent antitumor activity. Finally, recent studies have demonstrated that deviation of the immune response from Th2 to Th1 mediates potent anti-tumor response in otherwise resistant tumors (Hu et al. 1998). Thus, these findings suggest that IL-13 inhibitors or novel IL-13 antagonists may prove to be highly effective anticancer immunotherapeutics.

Targeting IL-13R for malignant brain tumor and RCC therapy

As human brain tumor and RCC cells express 15- to 500-fold higher numbers of high or intermediate affinity functional IL-13R compared to normal immune cells, IL-13 cytotoxin mediated efficient and specific cytotoxicity. Protein synthesis inhibition activity of IL-13 cytotoxin was confirmed by clonogenic assays that showed IL-13 cytotoxin can inhibit RCC cell colony formation in a dose-dependent manner (Puri et al. 1996). Interestingly, IL-13 cytotoxin has 10-fold lower binding affinity toward IL-13R on RCC cells (Puri et al. 1996). In HL-RCC cells, the value for native IL-13 binding was ~20 $\times$ 10^{-9} mol/l, compared with ~180 $\times$ 10^{-9} mol/l with IL-13 cytotoxin. This molecule was highly cytotoxic and cytotoxicity in RCC cell lines ranged between 0.03 to 17.5 ng/ml (Puri et al. 1996). These results suggested that if we produce a molecule that binds just as well as IL-13, we may have a much more cytotoxic agent.

Animal studies

As IL-13 cytotoxin mediated significant antitumor activity in tissue culture, we wished to examine the antitumor activity in an animal model of human disease. Various human tumor models including glioblastoma, AIDS-KS, and SCCHN were tested (Husain et al. 2001; Husain et al. 1997; Husain and Puri 2000; Kawakami et al. 2001b; Kawakami et al. 2000a). IL-13 cytotoxin mediated remarkable antitumor activity in these models when it was administered to subcutaneous tumor-bearing mice by different routes of administration. All mice in these models tolerated the therapy well without any visible signs of toxicity.

We also examined the antitumor activity of IL-13 cytotoxin in an animal model of human RCC as IL-13 cytotoxin was highly cytotoxic to human RCC cells *in vitro*. We attempted to generate a tumor model using various primary cultures of human RCC cell lines; however, after initial tumor growth, tumors spontaneously regressed in some animals. These observations hampered our efforts to test IL-13 cytotoxin *in vivo*. Therefore, we utilized matrigel, which helped development of RCC tumors when cells were injected subcutaneously in athymic nude mice. These tumors developed in 100% of animals, grew slowly, and did not regress spontaneously. IL-13 cytotoxin was tested in this model and various doses via different routes were administered to tumor-bearing hosts. IL-13 cytotoxin caused regression of these tumors in a dose-dependent manner. Intratumoral administration was the most effective route of administration followed by intraperitoneal (i.p.) administration. These responses were durable and 100% of the animals showed complete regression of their tumors versus 50% complete response in mice treated by the i.p. route (Husain and Puri, unpublished results). These results suggested that IL-13R-targeted IL-13 cytotoxin might also have antitumor activity in patients with renal cell carcinoma.

Toxicity and Pharmacokinetic studies

Pharmacokinetic and toxicological studies were performed for further development of IL-13 cytotoxin for clinical trial. Both male and female CD2F1 mice were administered various doses of IL-13 cytotoxin (i.v. or i.p., q.o.d.). These animals were examined for general toxicity, weight loss, serum chemistry, and hematology. In acute toxicity studies, hepatic enzymes, alanine aminotransferase (ALT), and aspartate aminotransferase (AST) were mildly elevated at 50 μg/kg doses. However, at 75 μg/kg doses, both AST and ALT were sharply elevated (three to six times higher than control mice). Interestingly, female mice showed a higher elevation of these enzymes compared to male mice. At lower doses (25 μg/kg or lower), no significant changes were observed either in AST or ALT. None of the treated mice showed hematological toxicity or vascular leak syndrome, a common side-effect seen in immunotoxin-based therapy. Finally, mice treated at the highest dose showed weight loss and three female mice out of five died at the 75 μg/kg dose. Autopsy analysis revealed moderate zonal hepatic necrosis (Husain et al. unpublished results). These results suggested that 50 μg/kg may be safely administered to these animals.

In a chronic toxicity study, a 50 μg/kg dose was administered i.v. QOD on alternate days for 9 days. At this dose and schedule, transient weight loss was observed more in female mice than male mice. Both transaminase values (AST and ALT) rose in female mice, but only AST rose in male mice. Serum alkaline phosphatase values were not increased in either male or female mice. Repeated injections of IL-13 cytotoxin at lower doses (12.5 and 25 μg/kg) did not induce any changes (Husain and Puri, unpublished results).

The molecular mechanism of IL-13 cytotoxin–induced hepatotoxicity is not clearly understood. Whether high levels of IL-13R are overexpressed in hepatocytes and IL-13 cytotoxin mediates cytotoxicity through these receptors or IL-13 cytotoxin induces cytotoxicity due to nonspecific uptake remains to be resolved. In any case, because human IL-13 may bind to murine IL-13R (Zurawski and de Vries 1994), the toxicity studies in mice may reflect human conditions indicating that IL-13 cytotoxin may be well tolerated in clinical trials at doses less than 50 μg/kg.

Toxicology and pharmacology studies were also performed in cynomolgus monkeys. IL-13 cytotoxin (12.5 or 50 μg/kg/day × 5 days) was administered daily for 5 days to cynomolgus monkeys of each sex and general clinical examination, serum chemistry, and hematology were performed at various time points. Similar to murine studies, these monkeys also showed transient dose-related elevation of hepatic transaminases, which reached maximum levels on days 5 to 8 of injection and subsequently returned to baseline at day 15 to 22. Interestingly, these monkeys showed a moderate decrease in serum cholesterol at the highest dose of IL-13 cytotoxin. Elevation of creatinine kinase was also noted on day 3, which subsequently declined to normal. The monkeys developed unusual dermatologic toxicity near the injection site at the highest dose. However, no hematological, renal toxicities, or vascular leak syndrome were observed. No monkey showed any bodyweight loss and IL-13 cytotoxin was well tolerated in cynomolgus monkeys.

The pharmacokinetic studies in CD2F1 mice, which received 50 μg/kg IL-13 cytotoxin i.v. showed a serum half life ($t_{1/2\beta}$) of the drug at 1.39 h with the C_{max} of 717 ng/ml, and the area under the curve (AUC) as 219 ng/h/ml. In monkey studies, $t_{1/2}$ was also short (Husain and Puri, unpublished results).

Clinical studies

Based on the preclinical profile of IL-13 cytotoxin, we initiated three different Phase I/II clinical trials in recurrent resectable malignant glioma patients (Lang et al. 2002; Prados et al. 2002a; Prados et al. 2002b; Prados et al. 2001; Weingart et al. 2001a; Weingart et al. 2001b). Patient inclusion criteria for the first study required a prior diagnosis of supratentorial malignant glioma, recurrence after radiotherapy (RT), measurable disease (1 to 5 cm in diameter), and a stereotaxic biopsy confirming malignant glioma at the study entry (Weingart et al. 2001a; Weingart et al. 2001b). In this clinical study, IL-13 cytotoxin is administered by micro-infusion via intratumoral catheters at 200 μl/catheter/h for 96 h (total 38.4 ml). The dose of the cytotoxin is increased in each cohort of three patients to determine the maximum tolerated dose (MTD). Cohorts are expanded to six patients if one dose limiting toxicity (DLT) is observed. At the initial dose level of 0.125 μg/ml infusion concentration, one of the first three patients developed a symptomatic increase in peritumoral edema soon after the first dose and was cleared by corticosteroids. Guidelines for steroid prophlylaxis were subsequently instituted and three additional patients were accrued prior to dose escalation. Dose escalation has proceeded in all three patient in each cohort at 0.25 and 0.5 μg/ml. A radiographic response was noted in one patient 7 weeks after the initial infusion of the drug. The histopathology of the resected tissues from this patient showed only necrosis. Two patients, who developed progressive neurologic findings and increased contrast enhancement, underwent debulking surgery. In both cases, histopathology of the resected tissue showed only necrosis. Prolonged patient survival has also been observed. A further dose-escalation study is currently underway to determine the MTD that will be used to formally evaluate efficacy of this cytotoxin in the Phase II portion of the study.

Because tumor cell-kill concentrations of IL-13 cytotoxin may be far below an MTD defined by neurotoxicity, preresection intra-tumoral administration may permit identification of a histologically effective concentration (HEC). Histologic efficacy is defined as geographic necrosis caused by loss of cellular integrity with eosinophilic staining or by complete cell loss. More than 90% of cell necrosis in the post-infusion specimen compared with the preinfusion biopsy would define histologic evidence of the drug. For these reasons, the second clinical trial (Lang et al. 2002; Prados et al. 2002a; Prados et al. 2002b; Prados et al. 2001) focused on determining the HEC of IL-13 cyto-toxin that may document the cytotoxicity to tumor. In stage 1 of the study, IL-13 cytotoxin is being infused for 2 days by IT continuous infusion into recurrent malignant glioma prior to surgical resection. In addition, we also aim to evaluate the toxicity of the drug infused via catheter into the brain adjacent to the tumor resection site after surgical resection. The secondary objective of this study also included the monitoring of disease progression and survival of the patients. Patients to be included in this trial must have had prior histologically confirmed diagnosis of supratentorial grade 3 or 4 malignant glioma including anaplastic astrocytoma, glioblastoma multi-forme, or malignant oligoastrocytoma. After biopsy and catheter placement on day 1, IL-13 cytotoxin is administered for 48 h at 400 μl/h on day 2 to 4. Tumor is resected on day 8, with the goal of en bloc removal, and tissue evaluated for necrosis adjacent to the catheter. Following resection, two or three catheters are placed into the brain adjacent to the operative site; on days 10 to 14, IL-13 cytotoxin is given by intracerebral micro-infusion (ICM) at 750 μl/h for 96 h. Escalating dose of IT and ICM concentrations, starting at 0.25 μg/ml, need to be studied separately in patients. After an HEC or MTD is identified, IT administration will be omitted in stage 2 of the study. Escalation of IL-13 cytotoxin concentration will then begin in ICM, and continue up to the documented HEC or MTD. Tumor specimens from two patients after IT injection at 0.5 μg/ml IL-13 cytotoxin revealed regional necrosis in an ovoid zone extending 1 to 2 cm from catheter tip, consistent with drug effect. Using ICM of 72 ml over 4 days (0.25 μg/ml), all patients tolerated infusion well and experienced no subsequent neurotoxicity. Patient accrual continues in the post-resection second stage of the study.

The third Phase I study focused on evaluating the MTD in terms of dura-tion of infusion and drug concentration of IL-13 cytotxin delivered by CED via one or two IT catheters into recurrent malignant glioma prior to surgical resection. Initially, cohorts of three to six patients received a fixed drug concentration (0.50 μg/ml) with escalating durations (4 to 7 days) of infusion to determine an MTD. After determination of maximum duration, drug concentration is escalated from 1.0 to 4.0 μg/ml in cohorts of three to six patients to determine an MTD based on concentration. Tumor necrosis is assessed 6 to 13 days after the end of infusion at the time of tumor resection. Seven patients have been included to study dose escalation. Patient accrual to determine the MTD based on duration continues. The safety and tolerabil-ity of IL-13 cytotoxin observed in this study correlates well with data from the other IL-13 cytotoxin studies in malignant glioma.

Similarly, our *in vitro* and preclinical results in RCC models encouraged us to initiate a Phase I clinical trial in patients with advanced renal cell carcinoma. Twelve patients (ten male, two female) with a median age of 64 years and metastatic disease, who had been previously treated with standard agents and had adequate organ function, were enrolled in this trial. The first cohort of three patients was dosed with IL-13 cytotoxin at 8 µg/kg i.v. bolus daily for five consecutive days. All three patients at this dose had reversible platelet consumption (decline >100 K) with rebound above baseline on day 8. One of three patients developed acute renal failure (ARF) requiring dialysis after five doses of IL-13 cytotoxin and the second patient developed ARF after three doses. This second patient did not require chronic dialysis. The renal biopsies from these two patients showed endothelial swelling, thrombotic glomerulopathy, and tubular necrosis. One of these two patients also developed transient grade 3 AST/ALT elevation and erythematous rash. However, the third patient did not show any renal toxicity. Upon further investigation, it was found that the third patient had preexisting antibodies to PE and therefore the IL-13 cytotoxin effect may have been neutralized (Kuzel et al. 2002).

Because of this unexpected renal toxicity, which was not observed in any animal studies, IL-13 cytotoxin doses were reduced to 1 or 2 µg/kg every alternate day for three doses. These seven patients did not show any renal or other toxicity indicating that 1 and 2 µg/kg dose is well tolerated. The next cohort of patients received 4 µg/kg dose of IL-13 cytotoxin every alternate day for 3 days. Adverse events in these patients noticed were grade 1/2 fever, chills, nausea, fatigue, edema, and headache, but no platelet consumption was observed. Of two patients treated at the 4 µg/kg dose, one had no nephrotoxicity; however the other patient, despite weakly positive preexisting antibody, had grade 4 ARF (irreversible) with marked platelet decrease (75% from baseline) and schistocytes on peripheral blood smear, unresponsive to plasmapheresis. Therefore, the MTD was established at 2 µg/kg × q.o.d.

Pharmacokinetic studies in anti-PE antibody-negative patients after 1, 2, 4, and 8 µg/kg doses on day 1 showed mean peak serum IL-13 cytotoxin levels of 4, 8, 13, and 79 ng/ml, respectively, with $t_{1/2\beta}$ of 30 min. This half-life is shorter compared to murine and monkey pharmacokinetic studies though the dose regimens were different. At 8 µg/kg, however, the mean peak level was 79 ng/ml with delayed clearance. All patients developed antibodies to IL-13 cytotoxin, and subsequent serum levels of the drug (if any) were markedly reduced.

The mechanism of IL-13 cytotoxin–induced ARF is not known. It is possible that ARF occurring in three of twelve patients is due to expression of IL-13 receptors in normal kidney cells or renal endothelium of these patients. As normal monkeys and mice did not show this toxicity and they did not have RCC, it is possible that the presence of RCC in kidney may activate IL-13R in normal kidney cells. Finally, since IL-13 cytoxin can eliminate 100% RCC cells *in vitro* at <1 ng/ml and the peak serum level of IL-13-PE at 2 µg/kg are well tolerated, it is possible that this dose level will mediate an antitumor effect. In addition, other schedules or routes of administration, e.g., intratumoral, may be tested that will avoid this toxic effect and may be beneficial to patients with RCC (Kuzel et al. 2002).

Conclusions and future directions

In this chapter, we have provided comprehensive information on the expression of IL-13 receptors on malignant brain tumor and RCC cell lines. Our studies have shown that the IL-13R is overexpressed in both types of tumor cell lines at mRNA and protein levels. Further, studies on the structure of IL-13R have demonstrated that human malignant brain tumor and RCC cells express type I IL-13R. We have also shown that a majority of these two cell lines express high numbers of IL-13R and are highly sensitive to the cytotoxic effect of IL-13 cytotoxin. However, cell lines that express low numbers of IL-13R were considerably less sensitive to this chimeric fusion cytotoxin. The molecular mechanism for differential expression of IL-13R on these tumor cell lines and cytotoxicity to IL-13 cytoxin is not clear. Our recent study has demonstrated that one of the two chains of IL-13R, IL-13Rα2 chain, binds IL-13 with higher affinity and is internalized after binding to ligand. In addition, we have demonstrated that gene transfer of this chain into tumor cells that do not express IL-13Rα2 sensitizes them to the cytotoxic effect of IL-13 cytotoxin (Kawakami et al. 2002d). These studies suggest that IL-13Rα2 chain improves sensitivity to IL-13 cytotoxin in renal cell carcinoma.

In vivo experiments have shown that the administration of IL-13 cytotoxin in a nude mouse xenograft tumor model can induce significant antitumor activity against human malignant gliomas, RCC, as well as other human cancers such as SSCHN and AIDS- KS, without any evidence of toxicity. Between these three different routes of administration, intratumoral IL-13 cytotoxin injection caused a complete eradication of the established human malignant gliomas and RCC tumors in less than a month. We expect that local injection of the toxin will most likely saturate the tumor bed, which will result in subsequent elimination of the tumor.

The Phase I clinical studies in recurrent glioma patients have demonstrated to date that intratumoral infusion of IL-13 cytotxin is well tolerated. The ongoing multi-center trials will further examine the safety and efficacy of IL-13 cytotoxin in the treatment of gliomas. As glioblastoma is a markedly heterogeneous tumor differing in receptor expression, a combination of therapy using cytotoxins and other anticancer drugs may be another candidate approach to overcome this deadly disease.

On the basis of our *in vitro* and *in vivo* results of RCC model, a Phase I safety and pharmacokinetic study was initiated and 12 patients with metastatic previously treated RCC were treated with IL-13 cytotoxin. This preliminary study established a 2 μg/kg dose as the MTD. A Phase II study involving a larger number of patients needs to be conducted at the MTD to determine the efficacy of this cytotoxic agent for human RCC and perhaps other tumors that express IL-13R. In addition, IL-13 cytotoxin should be tested with other known cytotoxic or immunomodulatory approaches for cancer therapy.

Acknowledgments

Clinical studies reported in this chapter for the therapy of renal cell carcinoma were performed in collaboration with Drs. Walter Urba, John Smith, and Bernard Fox of Earl Chile Cancer Institute, Portland, OR and Timothy Kuzel, Northwestern University Medical School, Chicago, IL; while preclinical studies were conducted in collaboration with Drs. Aquilur Rahman, Prafulla Gokhle, and Usha Kasid of Georgetown University, Washington, DC. We thank the New Approaches to Brain Tumor Therapy (NABTT), CNS Consortium, Baltimore MD, especially Drs. Stuart Grossman and Jon Weingart for their leadership in the first NCI-sponsored clinical trial. We also thank Drs. Michael Prados, Sandeep Kunwar, and Mitchel Berger of University of California San Francisco; Joseph Piepmeier of Yale University; Philip Gutin and Jeffrey Raizer of Memorial Sloan-Kettering, NY; Ken Aldape, Frederick Lang, and Alfred Yung of M.D. Anderson Cancer Center, TX for undertaking the second careful clinical study. The third trial is being conducted by Dr. Zvi Ram of the Chaim Sheba Medical Center in Israel; Drs. Gene Barnett and Michael Vogelbaum of the Cleveland Clinic; Dr. Shlomo Constantini of the Dana Hospital in Israel; Dr. Kevin Lillehei of the University of Colorado Health Sciences Center; Dr. Maximilian Medhorn of the University Hospital Liel in Germany; and Dr. Manfred Westphal of the University Hospital of Eppendorf in Germany. We are very thankful to all the collaborators. We also acknowledge Drs. Jeffrey Sherman, Imran Ahmad, and Jim Hussey of NeoPharm, Inc. for their support of the program. Many studies summarized in this chapter were performed as a part of the collaboration between the FDA and NeoPharm Inc. under a Cooperative Research and Development Agreement (CRADA).

Bibliography

Aman MJ, Tayebi N, Obiri NI, Puri RK, Modi WS, Leonard WJ. (1996). cDNA cloning and characterization of the human interleukin 13 receptor alpha chain. *J Biol Chem* 271:29265–29270.

Aversa G, Punnonen J, Cocks BG, de Waal Malefyt R, Vega F, Jr., Zurawski SM, Zurawski G, de Vries JE. (1993). An interleukin 4 (IL-4) mutant protein inhibits both IL-4 or IL-13-induced human immunoglobulin G4 (IgG4) and IgE synthesis and B cell proliferation: support for a common component shared by IL-4 and IL-13 receptors. *J Exp Med* 178:2213–2218.

Blease K, Jakubzick C, Schuh JM, Joshi BH, Puri RK, Hogaboam CM. (2001). IL-13 fusion cytotoxin ameliorates chronic fungal-induced allergic airway disease in mice. *J Immunol* 167:6583–6592.

Blease K, Schuh JM, Jakubzick C, Lukacs NW, Kunkel SL, Joshi BH, Puri RK, Kaplan MH, Hogaboam CM. (2002). Stat6-deficient mice develop airway hyperresponsiveness and peribronchial fibrosis during chronic fungal asthma. *Am J Pathol* 160:481–490.

Brown KD, Zurawski SM, Mosmann TR, Zurawski G. (1989). A family of small inducible proteins secreted by leukocytes are members of a new superfamily that includes leukocyte and fibroblast-derived inflammatory agents, growth factors, and indicators of various activation processes. *J Immunol* 142:679–687.

Campbell P, Marcus R. (2003). Monoclonal antibody therapy for lymphoma. *Blood Rev* 17:143–152.

Caput D, Laurent P, Kaghad M, Lelias JM, Lefort S, Vita N, Ferrara P. (1996). Cloning and characterization of a specific interleukin (IL)-13 binding protein structurally related to the IL-5 receptor alpha chain. *J Biol Chem* 271:16921–16926.

Chiaramonte MG, Cheever AW, Malley JD, Donaldson DD, Wynn TA. (2001). Studies of murine schistosomiasis reveal interleukin-13 blockade as a treatment for established and progressive liver fibrosis. *Hepatology* 34:273–282.

Chiaramonte MG, Donaldson DD, Cheever AW, Wynn TA. (1999a). An IL-13 inhibitor blocks the development of hepatic fibrosis during a T-helper type 2-dominated inflammatory response. *J Clin Invest* 104:777–785.

Chiaramonte MG, Mentink-Kane M, Jacobson BA, Cheever AW, Whitters MJ, Goad ME, Wong A, Collins M, Donaldson DD, Grusby MJ, Wynn TA. (2003). Regulation and function of the interleukin 13 receptor alpha 2 during a T helper cell type 2-dominant immune response. *J Exp Med* 197:687–701.

Chiaramonte MG, Schopf LR, Neben TY, Cheever AW, Donaldson DD, Wynn TA. (1999b). IL-13 is a key regulatory cytokine for Th2 cell-mediated pulmonary granuloma formation and IgE responses induced by *Schistosoma mansoni* eggs. *J Immunol* 162:920–930.

Cocks BG, de Waal Malefyt R, Galizzi JP, de Vries JE, Aversa G. (1993). IL-13 induces proliferation and differentiation of human B cells activated by the CD40 ligand. *Int Immunol* 5:657–663.

Colotta F, Re F, Muzio M, Bertini R, Polentarutti N, Sironi M, Giri JG, Dower SK, Sims JE, Mantovani A. (1993). Interleukin-1 type II receptor: a decoy target for IL-1 that is regulated by IL-4. *Science* 261:472–475.

Dabrowska A, Giermasz A, Golab J, Jakobisiak M. (2001). Potentiated antitumor effects of interleukin 12 and interferon alpha against B16F10 melanoma in mice. *Neoplasma* 48:358–361.

Danahay H, Atherton H, Jones G, Bridges RJ, Poll CT. (2002). Interleukin-13 induces a hypersecretory ion transport phenotype in human bronchial epithelial cells. *Am J Physiol Lung Cell Mol Physiol* 282:L226–236.

de Waal Malefyt R, Figdor CG, Huijbens R, Mohan-Peterson S, Bennett B, Culpepper J, Dang W, Zurawski G, de Vries JE. (1993). Effects of IL-13 on phenotype, cytokine production, and cytotoxic function of human monocytes. Comparison with IL-4 and modulation by IFN- gamma or IL-10. *J Immunol* 151:6370–6381.

Debinski W, Miner R, Leland P, Obiri NI, Puri RK. (1996). Receptor for interleukin (IL) 13 does not interact with IL4 but receptor for IL4 interacts with IL13 on human glioma cells. *J Biol Chem* 271:22428–22433.

Debinski W, Obiri NI, Pastan I, Puri RK. (1995a). A novel chimeric protein composed of interleukin 13 and Pseudomonas exotoxin is highly cytotoxic to human carcinoma cells expressing receptors for interleukin 13 and interleukin 4. *J Biol Chem* 270:16775–16780.

Debinski W, Obiri NI, Powers SK, Pastan I, Puri RK. (1995b). Human glioma cells overexpress receptors for interleukin 13 and are extremely sensitive to a novel chimeric protein composed of interleukin 13 and pseudomonas exotoxin. *Clin Cancer Res* 1:1253–1258.

Deen DF, Chiarodo A, Grimm EA, Fike JR, Israel MA, Kun LE, Levin VA, Marton LJ, Packer RJ, Pegg AE, et al. (1993). Brain Tumor Working Group Report on the 9th International Conference on Brain Tumor Research and Therapy. Organ System Program, National Cancer Institute. *J Neurooncol* 16:243–272.

Defrance T, Carayon P, Billian G, Guillemot JC, Minty A, Caput D, Ferrara P. (1994). Interleukin 13 is a B cell stimulating factor. *J Exp Med* 179:135–143.

Donaldson DD, Whitters MJ, Fitz LJ, Neben TY, Finnerty H, Henderson SL, O'Hara RM, Jr., Beier DR, Turner KJ, Wood CR, Collins M. (1998). The murine

IL-13 receptor alpha 2: molecular cloning, characterization, and comparison with murine IL-13 receptor alpha 1. *J Immunol* 161:2317–2324.

Fossa SD, Lien HH, Lindegaard M. (1991). Effect of recombinant interferon alfa on bone metastases of renal-cell carcinoma. *N Engl J Med* 324:633–634.

Frankel AE, Kreitman RJ, Sausville EA. (2000a). Targeted toxins. *Clin Cancer Res* 6:326–334.

Frankel AE, McCubrey JA, Miller MS, Delatte S, Ramage J, Kiser M, Kucera GL, Alexander RL, Beran M, Tagge EP, Kreitman RJ, Hogge DE. (2000b). Diphtheria toxin fused to human interleukin-3 is toxic to blasts from patients with myeloid leukemias. *Leukemia* 14:576–585.

Frankel AE, Powell BL, Hall PD, Case LD, Kreitman RJ. (2002). Phase I trial of a novel diphtheria toxin/granulocyte macrophage colony-stimulating factor fusion protein (DT388GMCSF) for refractory or relapsed acute myeloid leukemia. *Clin Cancer Res* 8:1004–1013.

Fyfe G, Fisher RI, Rosenberg SA, Sznol M, Parkinson DR, Louie AC. (1995). Results of treatment of 255 patients with metastatic renal cell carcinoma who received high-dose recombinant interleukin-2 therapy. *J Clin Oncol* 13:688–696.

Galanis E, Buckner JC, Dinapoli RP, Scheithauer BW, Jenkins RB, Wang CH, O'Fallon JR, Farr G, Jr. (1998). Clinical outcome of gliosarcoma compared with glioblastoma multiforme: North Central Cancer Treatment Group results. *J Neurosurg* 89:425–430.

Goldberg MR, Heimbrook DC, Russo P, Sarosdy MF, Greenberg RE, Giantonio BJ, Linehan WM, Walther M, Fisher HA, Messing E, et al. (1995). Phase I clinical study of the recombinant oncotoxin TP40 in superficial bladder cancer. *Clin Cancer Res* 1:57–61.

Greig NH, Ries LG, Yancik R, Rapoport SI. (1990). Increasing annual incidence of primary malignant brain tumors in the elderly. *J Natl Cancer Inst* 82:1621–1624.

Grossbard ML, Lambert JM, Goldmacher VS, Spector NL, Kinsella J, Eliseo L, Coral F, Taylor JA, Blattler WA, Epstein CL, et al. (1993). Anti-B4-blocked ricin: a phase I trial of 7-day continuous infusion in patients with B-cell neoplasms. *J Clin Oncol* 11:726–737.

Grunig G, Warnock M, Wakil AE, Venkayya R, Brombacher F, Rennick DM, Sheppard D, Mohrs M, Donaldson DD, Locksley RM, Corry DB. (1998). Requirement for IL-13 independently of IL-4 in experimental asthma. *Science* 282:2261–2263.

Guirguis LM, Yang JC, White DE, Steinberg SM, Liewehr DJ, Rosenberg SA, Schwartzentruber DJ. (2002). Safety and efficacy of high-dose interleukin-2 therapy in patients with brain metastases. *J Immunother* 25:82–87.

Hamann PR, Hinman LM, Beyer CF, Lindh D, Upeslacis J, Flowers DA, Bernstein I. (2002). An anti-CD33 antibody-calicheamicin conjugate for treatment of acute myeloid leukemia. Choice of linker. *Bioconjug Chem* 13:40–46.

Hilton DJ, Zhang JG, Metcalf D, Alexander WS, Nicola NA, Willson TA. (1996). Cloning and characterization of a binding subunit of the interleukin 13 receptor that is also a component of the interleukin 4 receptor. *Proc Natl Acad Sci USA* 93:497–501.

Hu HM, Urba WJ, Fox BA. (1998). Gene-modified tumor vaccine with therapeutic potential shifts tumor-specific T cell response from a type 2 to a type 1 cytokine profile. *J Immunol* 161:3033–3041.

Hurst R, White DE, Heiss J, Lee DS, Rosenberg SA, Schwartzentruber DJ. (1999). Brain metastasis after immunotherapy in patients with metastatic melanoma or renal cell cancer: is craniotomy indicated? *J Immunother* 22:356–362.

Husain SR, Behari N, Kreitman RJ, Pastan I, Puri RK. (1998). Complete regression of established human glioblastoma tumor xenograft by interleukin-4 toxin therapy. *Cancer Res* 58:3649–3653.

Husain SR, Joshi BH, Puri RK. (2001). Interleukin-13 receptor as a unique target for anti-glioblastoma therapy. *Int J Cancer* 92:168–175.

Husain SR, Obiri NI, Gill P, Zheng T, Pastan I, Debinski W, Puri RK. (1997). Receptor for interleukin 13 on AIDS-associated Kaposi's sarcoma cells serves as a new target for a potent Pseudomonas exotoxin-based chimeric toxin protein. *Clin Cancer Res* 3:151–156.

Husain SR, Puri RK. (2000). Interleukin-13 fusion cytotoxin as a potent targeted agent for AIDS-Kaposi's sarcoma xenograft. *Blood* 95:3506–3513.

Husain SR, Puri RK. (2003). Interleukin-13 receptor-directed cytotoxin for malignant glioma therapy: from bench to bedside. *J Neuro-Oncol* 65:37–48.

Idzerda RL, March CJ, Mosley B, Lyman SD, Bos TV, Girupel SD. (1990). Human interleukin 4 receptors confers biological responsiveness and defines a novel receptor superfamily. *J Exp Med* 171:861–873.

Iglewski BH, Kabat D. (1975). NAD-dependent inhibition of protein synthesis by Pseudomonas aeruginosa toxin. *Proc Natl Acad Sci USA* 72:2284–2288.

Indolfi P, Terenziani M, Casale F, Carli M, Bisogno G, Schiavetti A, Mancini A, Rondelli R, Pession A, Jenkner A, Pierani P, Tamaro P, De Bernardi B, Ferrari A, Santoro N, Giuliano M, Cecchetto G, Piva L, Surico G, Di Tullio MT. (2003). Renal cell carcinoma in children: a clinicopathologic study. *J Clin Oncol* 21:530–535.

Jakubzick C, Choi ES, Joshi BH, Keane MP, Kunkel SL, Puri RK, Hogaboam CM. (2003a). Therapeutic attenuation of pulmonary fibrosis via targeting of IL-4- and IL-13-responsive cells. *J Immunol* 171:2684–2693.

Jakubzick C, Choi ES, Kunkel SL, Joshi BH, Puri RK, Hogaboam CM. (2003b). Impact of interleukin-13 responsiveness on the synthetic and proliferative properties of Th1- and Th2-type pulmonary granuloma fibroblasts. *Am J Pathol* 162:1475–1486.

Jakubzick C, Kunkel SL, Joshi BH, Puri RK, Hogaboam CM. (2002). Interleukin-13 fusion cytotoxin arrests Schistosoma mansoni egg-induced pulmonary granuloma formation in mice. *Am J Pathol* 161:1283–1297.

Jemal A, Murray T, Samuels A, Ghafoor A, Ward E, Thun MJ. (2003). Cancer statistics, 2003. *CA Cancer J Clin* 53:5–26.

Joshi BH, Kawakami K, Leland P, Puri RK. (2002). Heterogeneity in interleukin-13 receptor expression and subunit structure in squamous cell carcinoma of head and neck: differential sensitivity to chimeric fusion proteins comprised of interleukin-13 and a mutated form of Pseudomonas exotoxin. *Clin Cancer Res* 8:1948–1956.

Joshi BH, Leland P, Puri RK. (2003). Identification and characterization of interleukin-13 receptor in human medulloblastoma and targeting these receptors with interleukin-13-pseudomonas exotoxin fusion protein. *Croat Med J* 44:455–462.

Joshi BH, Plautz GE, Puri RK. (2000). Interleukin-13 receptor alpha chain: a novel tumor-associated transmembrane protein in primary explants of human malignant gliomas. *Cancer Res* 60:1168–1172.

Joshi BH, Husain SR, and Puri RK (2000). Preclinical studies with IL-13PE38QQR for therapy of malignant glioma. *Drug News Perspect* 13: 599–605.

Kacha AK, Fallarino F, Markiewicz MA, Gajewski TF. (2000). Cutting edge: spontaneous rejection of poorly immunogenic P1.HTR tumors by Stat6-deficient mice. *J Immunol* 165:6024–6028

Kawakami K, Husain SR, Bright RK, Puri RK. (2001a) Gene transfer of interleukin 13 receptor alpha2 chain dramatically enhances the antitumor effect of IL-13 receptor-targeted cytotoxin in human prostate cancer xenografts. *Cancer Gene Ther* 8:861–868.

Kawakami K, Husain SR, Kawakami M, Puri RK. (2002a). Improved anti-tumor activity and safety of interleukin-13 receptor targeted cytotoxin by systemic continuous administration in head and neck cancer xenograft model. *Mol Med* 8:487–494.

Kawakami K, Joshi BH, Puri RK. (2000a). Sensitization of cancer cells to interleukin 13-pseudomonas exotoxin-induced cell death by gene transfer of interleukin 13 receptor alpha chain [in process citation]. *Hum Gene Ther* 11:1829–1835.

Kawakami K, Kawakami M, Joshi BH, Puri RK. (2001b). Interleukin-13 receptor-targeted cancer therapy in an immunodeficient animal model of human head and neck cancer. *Cancer Res* 61:6194–6200.

Kawakami K, Kawakami M, Puri RK. (2002b). IL-13 receptor-targeted cytotoxin cancer therapy leads to complete eradication of tumors with the aid of phagocytic cells in nude mice model of human cancer. *J Immunol* 169:7119–7126.

Kawakami K, Kawakami M, Snoy PJ, Husain SR, Puri RK. (2001c). *In vivo* overexpression of IL-13 receptor alpha2 chain inhibits tumorigenicity of human breast and pancreatic tumors in immunodeficient mice. *J Exp Med* 194:1743–1754.

Kawakami K, Taguchi J, Murata T, Puri RK. (2001d). The interleukin-13 receptor alpha2 chain: an essential component for binding and internalization but not for interleukin-13-induced signal transduction through the STAT6 pathway. *Blood* 97:2673–2679.

Kawakami K, Takeshita F, Puri RK. (2001e). Identification of distinct roles for a dileucine and a tyrosine internalization motif in the interleukin (IL)-13 binding component IL-13 receptor alpha 2 chain. *J Biol Chem* 276:25114–25120.

Kawakami M, Kawakami K, Puri RK. (2002c). Apoptotic pathways of cell death induced by an interleukin-13 receptor-targeted recombinant cytotoxin in head and neck cancer cells. *Cancer Immunol Immunother* 50:691–700.

Kawakami M, Kawakami K, Puri RK. (2002d). Intratumor administration of interleukin 13 receptor-targeted cytotoxin induces apoptotic cell death in human malignant glioma tumor xenografts. *Mol Cancer Ther* 1:999–1007.

Kawakami M, Leland P, Kawakami K, Puri RK. (2000b). Mutation and functional analysis of IL-13 receptors in human malignant glioma cells. *Oncol Res* 12:459–467.

Kawakami M, Kawakami K and Puri RK (2003). IL-4 Pseudomonas exotoxin fusion protein for malignant glioma therapy. *J Neuro-oncology* 65: 15–25.

Kohler I, Alliger P, Minty A, Caput D, Ferrara P, Holl-Neugebauer B, Rank G, Rieber EP. (1994). Human interleukin-13 activates the interleukin-4-dependent transcription factor NF-IL4 sharing a DNA binding motif with an interferon-gamma-induced nuclear binding factor. *FEBS Lett* 345:187–192.

Kohler I, Alliger P, Rieber EP. (1995). Activation of gene transcription by IL-4, IL-13 and IFN-gamma through a shared DNA binding motif. *Behring Inst Mitt* 78–86.

Kraft M, Lewis C, Pham D, Chu HW. (2001). IL-4, IL-13, and dexamethasone augment fibroblast proliferation in asthma. *J Allergy Clin Immunol* 107:602–606.

Kreitman RJ, Wilson WH, Robbins D, Margulies I, Stetler-Stevenson M, Waldmann TA, Pastan I. (1999). Responses in refractory hairy cell leukemia to a recombinant immunotoxin. *Blood* 94:3340–3348.

Kuzel T, Smith II J, Urba W, Fox B, Moudgil T, Strauss L, Joshi B, Puri R. (2002). IL13-PE38QQR cytototxind in advanced renal cell carcinoma (RCC): phase 1 and pharmacokinetic study. In *American Society of Clinical Oncology (ASCO)*, SanFrancisco, CA.

Kuznetsov VA, Puri RK. (1999). Kinetic analysis of high affinity forms of interleukin (IL)-13 receptors: suppression of IL-13 binding by IL-2 receptor gamma chain. *Biophys J* 77:154–172.

Lang F, Kunwar S, Strauss L, Piepmeier J, McDermott M, Fleming C, sherman J, Raizer J, Alalpe K, Yung WK, Husain SR, Chang SM, Berger M, Prados M, Puri RK. (2002). A clinical study of convection-enhanced delivery of IL-13PE38QQR cytotoxin pre- and post- resection of recurrent GBM. *American Society of Neuro-Oncologists*, Chicago, April 2002.

Lin JX, Migone TS, Tsang M, Friedmann M, Weatherbee JA, Zhou L, Yamauchi A, Bloom ET, Mietz J, John S, et al. (1995). The role of shared receptor motifs and common Stat proteins in the generation of cytokine pleiotropy and redundancy by IL-2, IL-4, IL-7, IL-13, and IL-15. *Immunity* 2:331–339.

Liu JY, Wei YQ, Yang L, Zhao X, Tian L, Hou JM, Niu T, Liu F, Jiang Y, Hu B, Wu Y, Su JM, Lou YY, He QM, Wen YJ, Yang JL, Kan B, Mao YQ, Luo F, Peng F. (2003). Immunotherapy of tumors with vaccine based on quail homologous vascular endothelial growth factor receptor-2. *Blood* 102:1815–1823.

Maini A, Hillman G, Haas GP, Wang CY, Montecillo E, Hamzavi F, Pontes JE, Leland P, Pastan I, Debinski W, Puri RK. (1997). Interleukin-13 receptors on human prostate carcinoma cell lines represent a novel target for a chimeric protein composed of IL-13 and a mutated form of Pseudomonas exotoxin. *J Urol* 158:948–953.

Matsuoka S, Tsurui H, Abe M, Terashima K, Nakamura K, Hamano Y, Ohtsuji M, Honma N, Serizawa I, Ishii Y, Takiguchi M, Hirose S, Shirai T. (2003). A monoclonal antibody to the alpha2 domain of murine major histocompatibility complex class I that specifically kills activated lymphocytes and blocks liver damage in the concanavalin A hepatitis model. *J Exp Med* 198:497–503.

McKenzie AN, Culpepper JA, de Waal Malefyt R, Briere F, Punnonen J, Aversa G, Sato A, Dang W, Cocks BG, Menon S, et al. (1993). Interleukin 13, a T-cell-derived cytokine that regulates human monocyte and B-cell function. *Proc Natl Acad Sci USA* 90:3735–3739.

Minty A, Chalon P, Derocq JM, Dumont X, Guillemot JC, Kaghad M, Labit C, Leplatois P, Liauzun P, Miloux B, et al. (1993). Interleukin-13 is a new human lymphokine regulating inflammatory and immune responses. *Nature* 362:248–250.

Morse B, Sypek JP, Donaldson DD, Haley KJ, Lilly CM. (2002). Effects of IL-13 on airway responses in the guinea pig. *Am J Physiol Lung Cell Mol Physiol* 282:L44–49.

Murata T, Husain SR, Mohri H, Puri RK. (1998a). Two different IL-13 receptor chains are expressed in normal human skin fibroblasts, and IL-4 and IL-13 mediate signal transduction through a common pathway. *Int Immunol* 10:1103–1110.

Murata T, Noguchi PD, Puri RK. (1995). Receptors for interleukin (IL)-4 do not associate with the common gamma chain, and IL-4 induces the phosphorylation of JAK2 tyrosine kinase in human colon carcinoma cells. *J Biol Chem* 270:30829–30836.

Murata T, Noguchi PD, Puri RK. (1996). IL-13 induces phosphorylation and activation of JAK2 Janus kinase in human colon carcinoma cell lines: similarities between IL-4 and IL-13 signaling. *J Immunol* 156:2972–2978.

Murata T, Obiri NI, Debinski W, Puri RK. (1997a). Structure of IL-13 receptor: analysis of subunit composition in cancer and immune cells. *Biochem Biophys Res Commun* 238:90–94.

Murata T, Obiri NI, Puri RK. (1997b). Human ovarian-carcinoma cell lines express IL-4 and IL-13 receptors: comparison between IL-4- and IL-13-induced signal transduction. *Int J Cancer* 70:230–240.

Murata T, Obiri NI, Puri RK. (1998b). Structure of and signal transduction through interleukin-4 and interleukin-13 receptors (review). *Int J Mol Med* 1:551–557.

Murata T, Puri RK. (1997). Comparison of IL-13- and IL-4-induced signaling in EBV-immortalized human B cells. *Cell Immunol* 175:33–40.

Murata T, Taguchi J, Puri RK. (1998c). Interleukin-13 receptor alpha' but not alpha chain: a functional component of interleukin-4 receptors. *Blood* 91:3884–3891.

Obiri NI, Debinski W, Leonard WJ, Puri RK. (1995). Receptor for interleukin 13. Interaction with interleukin 4 by a mechanism that does not involve the common gamma chain shared by receptors for interleukins 2, 4, 7, 9, and 15. *J Biol Chem* 270:8797–8804.

Obiri NI, Husain SR, Debinski W, Puri RK. (1996). Interleukin 13 inhibits growth of human renal cell carcinoma cells independently of the p140 interleukin 4 receptor chain. *Clin Cancer Res* 2:1743–1749.

Obiri NI, Leland P, Murata T, Debinski W, Puri RK. (1997a). The IL-13 receptor structure differs on various cell types and may share more than one component with IL-4 receptor. *J Immunol* 158:756–764.

Obiri NI, Murata T, Debinski W, Puri RK. (1997b). Modulation of interleukin (IL)-13 binding and signaling by the gamma c chain of the IL-2 receptor. *J Biol Chem* 272:20251–20258.

Oosterwijk E, Divgi CR, Brouwers A, Boerman OC, Larson SM, Mulders P, Old LJ. (2003). Monoclonal antibody-based therapy for renal cell carcinoma. *Urol Clin North Am* 30:623–631.

Ostrand-Rosenberg S, Grusby MJ, Clements VK. (2000). Cutting edge: STAT6-deficient mice have enhanced tumor immunity to primary and metastatic mammary carcinoma. *J Immunol* 165:6015–6019.

Pai LH, Wittes R, Setser A, Willingham MC, Pastan I. (1996). Treatment of advanced solid tumors with immunotoxin LMB-1: an antibody linked to Pseudomonas exotoxin. *Nat Med* 2:350–353.

Pai-Scherf LH, Villa J, Pearson D, Watson T, Liu E, Willingham MC, Pastan I. (1999). Hepatotoxicity in cancer patients receiving erb-38, a recombinant immunotoxin that targets the erbB2 receptor. *Clin Cancer Res* 5:2311–2315.

Pastan I, Chaudhary V, FitzGerald DJ. (1992). Recombinant toxins as novel therapeutic agents. *Annu Rev Biochem* 61:331–354.

Pastan I, Pai LH, Brinkmann U, FitzGerald DI. (1995). Recombinant toxins:new therapeutic agents for cancer. *Ann N Y Acad Sci* 758:345–354.

Prados M, Lang F, sherman J, Strauss L, Fleming C, Alalpe K, Kunwar S, Yung WK, Chang SM, Husain SR, Gutin P, Raizer J, Piepmeier J, Berger M, McDermott M, Puri RK. (2002a). Convection-Enhanced Delivery (CED) by Positive Pressure Infusion for Intra-tumoral and Peri-tumoral Administration of IL-13PE38QQR a recombinant Tumor-Targeted cytotoxin in Recurrent Malignant Glioma. *J Neuro-Oncol* 4:S78.

Prados M, Lang F, Strauss L, Fleming C, Alalpe K, Kunwar S, Yung WK, Chang SM, Husain SR, Gutin P, Raizer J, Piepmeier J, Berger M, McDermott M, Puri RK. (2002b). Intrtumoral and intracerebral microinfusion of IL13-PE38QQR cytotoxin: Phase I/II study of pre- and post-resection infusions in recurrent resectable malignant glioma. *Am Soc Clin Oncol* 21:69B.

Prados M, Lang F, Strauss L, Fleming C, Aldape K, Kunwar S, Yung WKA, Husain SR, Chang SM, Gutin P, Raizer J, Piepmeier J, Berger M, Puri RK. (2001). Pre and post-resection interstitial infusions of IL13-PE38QQR cytotoxin: Phase I study in recurrent respectable malignant glioma. In *First Qudrennial Meeting — World Federation of Neuro-Oncology*, Washington, D.C.

Punnonen J, Aversa G, Cocks BG, McKenzie AN, Menon S, Zurawski G, de Waal Malefyt R, de Vries JE. (1993). Interleukin 13 induces interleukin 4-independent IgG4 and IgE synthesis and CD23 expression by human B cells. *Proc Natl Acad Sci USA* 90:3730–3734.

Puri RK, Leland P, Obiri NI, Husain SR, Kreitman RJ, Haas GP, Pastan I, Debinski W. (1996). Targeting of interleukin-13 receptor on human renal cell carcinoma cells by a recombinant chimeric protein composed of interleukin-13 and a truncated form of Pseudomonas exotoxin A (PE38QQR). *Blood* 87:4333–4339.

Puri, RK.(2000). Cytotoxins directed at interleukin-4 receptors as therapy for human brain tumors. Methods in Molecular Biology, Vol. 166. In *Immunotoxin Methods and Protocols*. WA Hall (ed.). Human Press, Totowa, NJ. 155–176.

Rahaman SO, Sharma P, Harbor PC, Aman MJ, Vogelbaum MA, Haque SJ. (2002). IL-13R(alpha)2, a decoy receptor for IL-13 acts as an inhibitor of IL-4-dependent signal transduction in glioblastoma cells. *Cancer Res* 62:1103–1109.

Ries LAG, Eisner MP, Kosary CL, Hankey BF, Miller BA, Clegg LX, Edwards BK. (2000). *SEERS Cancer Statistics Review 1973–1997*. National Cancer Institute, Bethesda, MD.

Rosenberg SA. (1997). Keynote address: perspectives on the use of interleukin-2 in cancer treatment. *Cancer J Sci Am* 3 Suppl 1:S2–6.

Scappaticci FA. (2002). Mechanisms and future directions for angiogenesis-based cancer therapies. *J Clin Oncol* 20:3906–3927.

Sela B. (1999). [Interleukin IL-13: a central mediator in allergic asthma]. *Harefuah* 137:317–319.

Smerz-Bertling C, Duschl A. (1995). Both interleukin 4 and interleukin 13 induce tyrosine phosphorylation of the 140-kDa subunit of the interleukin 4 receptor. *J Biol Chem* 270:966–970.

States NCfHSftU. (1994). National Center for Health Statistics for the United States:1997.

Terabe M, Matsui S, Noben-Trauth N, Chen H, Watson C, Donaldson DD, Carbone DP, Paul WE, Berzofsky JA. (2000). NKT cell-mediated repression of tumor immunosurveillance by IL-13 and the IL-4R-STAT6 pathway. *Nat Immunol* 1:515–520.

Tomkinson A, Duez C, Cieslewicz G, Pratt JC, Joetham A, Shanafelt MC, Gundel R, Gelfand EW. (2001). A murine IL-4 receptor antagonist that inhibits IL-4- and IL-13-induced responses prevents antigen-induced airway eosinophilia and airway hyperresponsiveness. *J Immunol* 166:5792–5800.

Valery CA, Seilhean D, Boyer O, Marro B, Hauw JJ, Kemeny JL, Marsault C, Philippon J, Klatzmann D. (2002). Long-term survival after gene therapy for a recurrent glioblastoma. *Neurology* 58:1109–1112.

van Der Velden VH, te Marvelde JG, Hoogeveen PG, Bernstein ID, Houtsmuller AB, Berger MS, van Dongen JJ. (2001). Targeting of the CD33-calicheamicin immunoconjugate Mylotarg (CMA-676) in acute myeloid leukemia: *in vivo* and *in vitro* saturation and internalization by leukemic and normal myeloid cells. *Blood* 97:3197–3204.

Vita N, Lefort S, Laurent P, Caput D, Ferrara P. (1995). Characterization and comparison of the interleukin 13 receptor with the interleukin 4 receptor on several cell types. *J Biol Chem* 270:3512–3517.

Vogel G. (1998a). Interleukin-13's key role in asthma shown. *Science* 282:2168.

Vogel G. (1998b). Interleukin-13's key role in asthma shown [news; comment]. *Science* 282:2168.

Wagner JR, Walther MM, Linehan WM, White DE, Rosenberg SA, Yang JC. (1999). Interleukin-2 based immunotherapy for metastatic renal cell carcinoma with the kidney in place. *J Urol* 162:43–45.

Waldmann TA, Pastan IH, Gansow OA, Junghans RP. (1992). The multichain interleukin-2 receptor: a target for immunotherapy. *Ann Intern Med* 116:148–160.

Weingart J, Grossman SA, Bohan E, Fisher JD, Strauss L, Puri RK. (2001). Phase I/II study of interstitial infusion of IL13-PE38QQR cytotoxin in recurrent malignant glioma. In: *First Qudrennial meeting — World Federation of Neuro-Oncology*, Washington, D.C., A-12.

Weingart J, Strauss L, Grossman SA, Markett J, Tatter S, Fisher JD, Fleming C, Puri RK. (2002). Phase I/II study: intra-tumoral infusion of IL13-PE38QQR cytotoxin for recurrent supratentorial malignat glioma. *Neuro-Oncol* 4:379.

Welham MJ, Learmonth L, Bone H, Schrader JW. (1995). Interleukin-13 signal transduction in lymphohemopoietic cells. Similarities and differences in signal transduction with interleukin-4 and insulin. *J Biol Chem* 270:12286–12296.

Wills-Karp M, Luyimbazi J, Xu X, Schofield B, Neben TY, Karp CL, Donaldson DD. (1998). Interleukin-13: central mediator of allergic asthma. *Science* 282: 2258–2261.

Wood N, Whitters MJ, Jacobson BA, Witek J, Sypek JP, Kasaian M, Eppihimer MJ, Unger M, Tanaka T, Goldman SJ, Collins M, Donaldson DD, Grusby MJ. (2003). Enhanced interleukin (IL)-13 responses in mice lacking IL-13 receptor alpha 2. *J Exp Med* 197:703–709.

Zhang JG, Hilton DJ, Willson TA, McFarlane C, Roberts BA, Moritz RL, Simpson RJ, Alexander WS, Metcalf D, Nicola NA. (1997). Identification, purification, and characterization of a soluble interleukin (IL)-13-binding protein. Evidence that it is distinct from the cloned Il-13 receptor and Il-4 receptor alpha-chains. *J Biol Chem* 272:9474–9480.

Zimmermann N, Hershey GK, Foster PS, Rothenberg ME. (2003). Chemokines in asthma: cooperative interaction between chemokines and IL-13. *J Allergy Clin Immunol* 111:227–242; quiz 243.

Zucchi A, Mearini L, Mearini E, Costantini E. (2003). Stage pT1 renal cell carcinoma: review of the prognostic significance of size. *Urol Int* 70:47–50.

Zurawski G, de Vries JE. (1994). Interleukin 13, an interleukin 4-like cytokine that acts on monocytes and B cells, but not on T cells. *Immunol Today* 15:19–26.

Zurawski SM, Chomarat P, Djossou O, Bidaud C, McKenzie AN, Miossec P, Banchereau J, Zurawski G. (1995). The primary binding subunit of the human interleukin-4 receptor is also a component of the interleukin-13 receptor. *J Biol Chem* 270:13869–13878.

Zurawski SM, Vega F, Jr., Huyghe B, Zurawski G. (1993). Receptors for interleukin-13 and interleukin-4 are complex and share a novel component that functions in signal transduction. *Embo J* 12:2663–2670.

part two

Ricin-based immunotoxins/ cytotoxins

chapter four

Ricin immunotoxins in lymphomas: Clinical application

Roland Schnell and Andreas Engert

Contents

Introduction

Although most patients with Hodgkin's lymphoma (HL) and many patients with non-Hodgkin's lymphoma (NHL) can be cured by conventional

0-4152-6365-4/05/$0.00+$1.50
© 2005 by CRC Press

modalities including radiotherapy and polychemotherapy, a substantial proportion of patients will succumb to their disease. The major reason for tumor recurrence is the development of cell clones, which are resistant to conventional therapy. These cells might be eradicated by new immunotherapeutic agents such as monoclonal antibodies (MAbs) or MAb-based constructs. Many naked murine MAbs demonstrated only moderate or no antitumor activity. Thus, various immunotherapeutic approaches were investigated including immunotoxins (ITs), radioimmunoconjugates or bispecific antibodies to enhance cytotoxicity. ITs are hybrid molecules constructed by linking or fusing a MAb or other cell-binding ligands to toxins or subunits of plantal, bacterial, or fungal origin. Ricin, which is derived from the seeds of *Ricinus communis* (castor bean), is the most widely used toxin for chemically linked ITs. Over the past decade several phase I/II trials with ricin-based ITs were conducted in patients with refractory NHL and HL. In this chapter we describe the clinical application of ricin-based ITs.

Background

Most patients with relapsed lymphoma are incurable with conventional treatment. High-dose chemotherapy (HDCT) followed by transplantation of autologous bone marrow (ABMT) or peripheral blood stem cell (PSCT) is currently being used to treat patients with relapsed NHL and HL. However, HDCT can be applied only to a minority of patients (i.e., those younger than 65, and those with no major organ dysfunction). Furthermore, substantial morbidity and potential mortality is associated with these aggressive regimens. Data from lymphoma patients as well as from patients with other malignant diseases, including colorectal cancer and myeloid leukemia indicate that residual tumor cells remaining after first-line treatment can cause late relapses (Gribben et al., 1991; Gribben et al., 1993; Kanzler et al., 1996; Riethmuller et al., 1994; Wolf et al., 1996; Roy et al., 1991). Thus, eliminating residual lymphoma cells after first-line treatment might reduce relapse rates and improve the outcome in patients with malignant lymphoma. Before the advent of effective "naked" MAbs against Hodgkin/Reed-Sternberg (H-RS) (Borchmann et al., 2003; Wahl et al., 2002), and lymphoma cells (Maloney, 1999; Leonard and Link, 2002), ITs consisting of a specific cell-binding moiety and a potent toxin subunit were constructed to destroy selectively these malignant cells (Vitetta et al., 1987). For several reasons lymphoma is a very promising disease for IT treatment: First, lymphoma cells express surface antigens such as CD19, CD22, CD25, and CD30 (Anderson et al., 1984; Agnarsson and Kadin, 1989; Stein et al., 1985), which are internalized after binding the IT. These antigens are present only on normal human lymphoma cells but not on stem cells and other organs. Second, the mechanism of cell destruction by ITs is different from that of conventional agents, thus circumventing drug resistance. Third, ITs are capable of killing dormant non-dividing cells.

Ricin

Ricin was found by Stillmark in 1889 (Stillmark, 1897) as the first plant lectin from the seeds of the castor plant, *Ricinus communis*. Ricin belongs to the class of ribosome inhibiting proteins (RIPs). These toxins can be classified into type-1 RIPs (saporin), which are single-chain proteins, and type-2 RIPs (ricin) containing two polypeptide chains. The A-chain (30 kDa), conferring RNA N-glycosidase activity to cleave a specific adenine base from ribosomal RNA, is connected through a disulfide linkage with the B-chain (32 kDa). Ricin acts as a galactose-specific lectin (Youle et al., 1981) binding to normal mammalian cell-surface oligosaccharides (Simmons et al., 1986). Bound ricin is internalized by receptor-mediated endocytosis and by clathrin-independent endocytosis. Subsequently, ricin is delivered to a transferring receptor-enriched endosomal compartment. Approximately 90 to 95% of the internalized toxin is recycled to the cell surface or delivered to the lysosomes. The remaining toxin is delivered from the endosome to the trans-Golgi network. An increasing acidification induces conformational changes rendering the toxin sensitive to sequence-specific proteolysis. The interchain disulfide-bond is reduced before the catalytic A-chain domain is delivered to the ribosomes where it inactivates the 60S ribosomal subunit needed for the binding of elongation factor-2 during protein synthesis. The enzymatic A-chain hydrolyzes an adenin–ribose linkage within the 28S rRNA whereby inhibiting the protein synthesis (Endo and Tsurugi, 1988; Endo et al., 1987; Endo and Tsurugi, 1987).

Ricin A-chain ITs

Coupling of unmodified native ricin to MAbs resulted in substantial unspecific toxicity against nontarget cells (Vitetta et al., 1993). Subsequently, only the enzymatically active ricin A-chain was used for construction of ricin A-chain ITs replacing the lectin-binding B-chain of the whole ricin molecule with MAbs or MAb fragments. Thus, ricin A-chain ITs bind selectively to their specific target cells, which are subsequently destroyed upon internalization. The first generation of ricin A-chain ITs defined as a construct of intact MAb attached via nonsterically hindered linkers such as N-succinidyl-3-(2-pyridyldithio) propionate (SPDP) to native ricin A-chain were active *in vitro* but showed poor efficacy when used *in vivo* (Krolick et al., 1982). Thus, second-generation ricin A-chain ITs were developed based on three key modifications: (1) The constructs were highly purified via blue sepharose (Knowles and Thorpe, 1987) separating unreacted material from the IT. (2) Deglycosylation of mannose and fucose residues of the A-chain (dgA) reduced unspecific binding to liver and RES cells (Thorpe et al., 1985; Thorpe et al., 1987). (3) The introduction of the new crosslinker 4-succinimidyl-oxycarbonyl-α-methyl-α-(2-pyridyldithio) toluene (SMPT) with a sterically hindered disulfide bond between MAb and toxin (Thorpe et al., 1987) improved the stability (Figure 4.1). These ITs showed a greatly improved half-life of up to eight-fold *in vivo* (Blakey et al., 1987) and produced

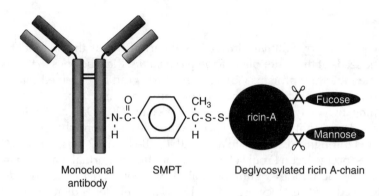

Figure 4.1   Model of a second-generation ricin A-chain immunotoxin. The toxic unit is bound via a sterically hindered disulfide linker (SMPT) to the binding unit. The deglycosylation of the A-chain abolished binding by mannose and fucose receptors on liver cells and cells of the reticulo-histiocytic system. These modifications extended half-life *in vivo* and decreased unspecific toxicity

significantly better antitumor effects than their predecessors in animal models (Fulton et al., 1988).

Blocked ricin ITs

Ricin A-chain ITs vary considerably in their potency. Only about 25% of MAb conjugated to ricin A-chain form ITs with acceptable cytotoxicity (Vitetta and Thorpe, 1991). Various attempts were evaluated to utilize the translocation function of the B-chain either in its single form or linked to MAbs to potentiate A-chain ITs (Vitetta et al., 1983; Vitetta et al., 1984). Alternatively, the whole ricin molecule was used as toxic moiety with preservation of the transport function of the B-chain, which is covalently modified at galactose-binding sites to subvert the nonspecific binding (Lambert et al., 1991b). Therefore, modified glycopeptides containing N-linked oligosaccharides derived from fetuin were constructed with terminal galactose residues available for binding and a reactive diclorotriazine moiety available for cross-linking to a protein. These ligands adhere to the galactose binding sites and then become covalently linked to the B-chain thereby minimizing the binding capability of ricin. This "blocked ricin" demonstrates a 1000-fold reduction of *in vitro* potency against a human B-cell NHL cell line (Namalwa) compared with unmodified ricin (Lambert et al., 1991a). In contrast, the full toxicity potential of native ricin is restored when blocked ricin is linked to a B-cell–specific MAb.

Ricin A-chain ITs in NHL

Several clinical phase I and II trials with ricin A-chain ITs of the first and second generation against target antigens such as CD5, CD19, and CD22 were conducted in patients with T-cell and B-cell derived NHL (Table 4.1).

Table 4.1 Clinical trials with ricin A-chain ITs in NHL patients

Disease	Antigen	Construct	Application	Toxicity (MTD)	Immunogenicity	Response	Reference
CTCL	CD5	H65-RTA	1-h infusion day 1–10	VLS, hepatotoxicity (0.33 mg/kg/day)	10/12 anti-immuno-conjugate response	4/14 PR	(LeMaistre et al., 1991)
B-NHL	CD22	Fab' RFB4-dgA	4-h infusion day 1–3–5–7	VLS, myalgia, rhabdomyolysis (75 mg/m²)	1/14 HAMA 4/14 HARA	5/14 PR	(Vitetta et al., 1991)
B-NHL	CD22	RFB4-SMPT-dgA	4-h infusion day 1–3–5–7	VLS, myalgia, rhabdomyolysis (32 mg/m²)	7/26 HAMA 8/26 HARA	1/26 CR, 5/26 PR	(Amlot et al., 1993)
B-NHL	CD22	RFB4-SMPT-dgA	Continuous infusion Day 1–8	VLS (19 mg/m²)	5/15 HAMA 6/15 HARA	4/16 PR	(Sausville et al., 1995)
B-NHL	CD19	IgG-HD37-dgA	4-h infusion day 1–3–5–7	VLS, aphasia, rhabdomyolysis, acrocyanosis (16 mg/m²)	4/15 HAMA 5/15 HARA	1/23 CR, 1/23 PR	(Stone et al., 1996)
B-NHL	CD19	IgG-HD37-dgA	Continuous infusion Day 1–8	VLS, acrocyanosis (19 mg/m²)	2/8 HAMA 2/8 HARA	1/9 PR	(Stone et al., 1996)
B-NHL	CD19	IgG-HD37-dgA	4-h infusion day 1–3–5–7	VLS, myalgia, rhabdomyolysis (16 mg/m²)	2/7 HAMA/HARA	1/8 PR	(Conry et al., 1995)
B-NHL	CD19 CD22	RFB4-SMPT-dgA IgG-HD37-dgA	Continuous infusion Day 1–8	VLs, hemolytic uremic syndrome	5/22 HAMA 1/22 HARA (10 mg/m²)	2/22 PR	(Messmann et al., 2000)

Abbreviations: NHL, non-Hodgkin's lymphoma; VLS, vascular leak syndrome, HAMA, human anti-mouse antibodies; HARA, human anti-ricin antibodies; CR, complete remission; PR, partial remission.

Ricin A-chain ITs in NHL against CD5

An initial trial with a first-generation ricin A-chain IT in NHL enrolled 14 patients with cutaneous T-cell lymphoma (LeMaistre et al., 1991). The anti-CD5 IT, H65-RTA, consisted of the murine MAb H65 linked to undeglycosylated ricin A-chain. The CD5 antigen is expressed on cutaneous T-cell lymphoma cells and on lymphocytes but not on other normal human cells (Wood et al., 1982; Holden et al., 1982). Patients received up to three cycles of H65-RTA. The maximum tolerated dose (MTD) of H65-RTA was defined at 0.33 mg/kg/day administered intravenously over 1 h for 10 days. The dominant toxicity was the occurrence of reversible hepatotoxicity and vascular leak syndrome (VLS), characterized by hypoalbuminemia, edema, weight gain, hypotension, tachycardia, dyspnea, and myalgia. Peak serum concentrations were dose-dependent, ranging from 1.13 to 5.56 μg/ml, with a terminal half-life of 1.0 to 2.9 h. The development of human anti-mouse antibodies (HAMA) and human anti-ricin antibodies (HARA) against the immunoconjugate was associated with a lower peak drug level, but not with enhanced side effects. In four patients a partial response (PR) occurred lasting from 3 to 8 months. This IT was also used in patients with steroid-refractory graft-versus-host reaction after allogeneic bone marrow transplantation resulting in 50% durable complete or partial responses (Byers et al., 1990).

Ricin A-chain ITs in NHL against CD22

The CD22 antigen is expressed early in the B-cell lineage, shortly after CD19 at the late pro-B-cell stage, although mainly as an intracellular protein (Banchereau and Rousset, 1992). Later in the B-cell development, CD22 is strongly expressed on the cell surface. Vitetta and colleagues have conducted a number of different clinical trials using the anti-CD22 MAb RFB4 or the RFB4 Fab'-fragment linked to dgA (Vitetta et al., 1991; Amlot et al., 1993). In the initial clinical study the RFB4 Fab'-dgA IT was used in 15 patients with low- (6), intermediate- (8), and high-grade (1) lymphoma. The patients received a 4 h i.v. infusion every other day over a 7 day period. As can be expected for a relatively small molecule the serum half-life was only 86 min. Serum levels were undetectable within 8 to 12 h after start of the infusion. The MTD was 75 mg/m^2 and dose-limiting toxicities (DLT) included VLS, fever, myalgia, rhabdomyolysis, and expressive aphasia. Thirty percent of evaluable patients developed HARA and one patient HAMA. In 5 of 14 patients a short (lasting 5 weeks to 4 months) PR occurred.

Since the production of Fab ITs was eight times more expensive when compared to IgG-ITs, whole MAb constructs were used instead of Fab-fragments in subsequent trials. In a phase I study using the bolus administration (four infusions 48 h apart), 26 patients with relapsed B-NHL were treated with the IgG counterpart RFB4-SMPT-dgA (Amlot et al., 1993). The MTD of RFB4-SMPT-dgA was defined at 32 mg/m^2 by VLS. Half-life ($T_{1/2}$)of the construct did not correlate directly with the given dose and averaged 7.8 h. Compared to the Fab'-IT, the toxicity profile was similar but there was a trend toward decreased toxicity in patients with evidence of bulky disease. HAMAs were observed more frequently (7 of 26 patients). One complete remission (CR) and five PRs were reported. In this trial high-peak concentrations of the IT, a longer $T_{1/2}$, and large area under the curve (AUC) correlated with both toxicity and clinical response. In a subsequent phase I trial the same IT was administered as continuous infusion over 8 days to reduce the incidence of VLS (Sausville et al., 1995). Nevertheless, compared to the intermittent bolus regimen this study gave similar profiles in terms of toxicity, immunogenicity,y and clinical response. The MTD was reached at 19 mg/m². High IT serum concentrations (>1000 ng/ml) by day 3 and the absence of circulating tumor cells were predictive for severe VLS. Even small numbers of circulating tumor cells (detectable by flow cytometry) prevented from severe VLS. Of 16 evaluable patients 4 achieved a PR. In all trials with this anti-CD22 IT, there was a notable trend to better clinical response in patients with small tumor burden (<100 cm²). In summary, there is no advantage of continuous infusion compared with the bolus regimen. However, bolus infusion is less expensive and more convenient. An obstacle in the use of anti-CD22-directed ITs is the variable expression of CD22 with a maximum of 70% positive tumor cells (Engert et al., 1998; Anderson et al., 1984). Thus, pan B-cell markers like CD19 or CD20 might be more reasonable targets.

Ricin A-chain ITs in NHL against CD19

The CD19 antigen is one of the earliest known B-cell markers and appears at the pro-B-cell stage (Banchereau and Rousset, 1992). CD19 is expressed in more than 95% of B-NHL cases (Anderson et al., 1984). The anti-CD19 dgA IT HD37-dgA was administered in phase I trials either using intermittent bolus or continuous regimen. Eight patients with refractory B-NHL were treated with four doses at 4 h intervals with total doses ranging from 4 to 12 mg/m² (Conry et al., 1995). The plasma half-life of HD37-dgA averaged 17 h and peak serum concentrations varied from 0.36 to 5.63 μg/ml. The toxicity profile was mainly related to VLS with one patient dying from severe VLS with bronchopneumonia and rhabdomyolysis. The MTD was established at 8 mg/m². Two of seven evaluable patients developed HAMA and HARA. None of the patients responded to treatment. In another trial, Stone and colleagues (Stone et al., 1996) treated 32 patients with low- or intermediate-grade B-NHL with the same IT. Twenty-three patients received the bolus schedule and a further nine patients the continuous infusion. When given as intermittent bolus infusion the MTD was 16 mg/m² with the DLTs consisting of VLS, aphasia, and rhabdomyolysis at 24 mg/m². Using the continuous infusion regimen, the MTD was again defined by VLS at 19.2 mg/m², which is comparable with the bolus administration. At the MTD of both regimens, 10% of patients experienced acrocyanosis with reversible superficial distal digital skin necrosis in the absence of overt evidence of systemic vasculitis. This phenomenon was not seen earlier, during IT treatment and the etiology remains unclear. Compared to the trial of Conry (Conry et al., 1995) less toxicity was observed probably related to the different intervals of IT administration. Of 23 evaluable patients on the bolus schedule, there was 1 persisting CR (40+ months) and 1 PR. Of 9 evaluable patients on the continuous infusion regimen, there was 1 PR. Pharmacokinetic parameters for the bolus regimen at the MTD showed a mean maximum serum concentration of 1209 ng/ml, with a median $T_{1/2}$ for all courses of 18.2 h. For the CI regimen at MTD, the mean maximum serum concentration was 963 ng/ml, with a median $T_{1/2}$ for all courses of 22.8 h. There was a clear linear correlation between the maximum serum concentration of the IT and the administered dose. Twenty-five percent of the patients on the bolus infusion regimen and 30% on the continuous infusion regimen developed HAMA and/or HARA. Again, there is no significant advantage for either of the regimens.

Ricin A-chain IT cocktail in NHL

One obstacle associated with MAb-based therapeutic approaches is the heterogenicity of tumor cells in malignant lymphoma (Anderson et al., 1984). A further problem observed after IT treatment in animal models is the survival of antigen-deficient mutants causing late relapse (Thorpe et al., 1988). A "cocktail" of two or more ITs against different antigens might overcome this problem. Preclinical studies in SCID mice with Daudi lymphoma showed superior results with a mixture of the anti-CD22 IT RFB4-SMPT-dgA and the anti-CD19 IT HD37-SMPT-dgA as compared to either IT alone (Ghetie et al., 1992). This "cocktail" was curative in SCID mice with advanced Daudi lymphoma after initial chemotherapy (Ghetie et al., 1996). Similar observations were made in nude mice with solid Hodgkin tumors receiving either a combination of two dgA-ITs directed against CD25 and IRac or the single ITs (Engert et al., 1995).

Thus, 22 patients with refractory B-NHL were treated in a phase I trial with a continuous infusion of a combination (mixture 1:1) of IgG-HD37.dgA and IgG-RFB4.dgA (Combotox) at doses of 10 to 30 mg/m^2 (Messmann et al., 2000). The MTD was reached at 10 mg/m^2. Toxicities in this trial included VLS and hemolytic uremic syndrome (HUS). Two patients died related to HUS. Patients with more than 49/µl circulating tumor cells in peripheral blood tolerated all doses without major toxicity. The maximum level of serum IT achieved in this group was 345 ng/ml of RFB4-dgA and 660 ng/ml of HD37-dgA, respectively. In contrast, patients with less circulating tumor cells had unpredictable clinical courses. Prior autologous bone marrow or peripheral blood stem cell transplantation ($P_2 = 0.003$) and a history of radiation therapy ($P_2 = 0.036$) were associated with significant mortality. Since both, high-dose chemotherapy and irradiation, can damage the endothelium (Baker and Krochak, 1989; Bertomeu et al., 1990), this might have predisposed for IT-mediated VLS (Lindstrom et al., 1997; Kuan et al., 1995; Soler-Rodriguez et al., 1993). In addition, thawed HD37-dgA tended to aggregate and formed HD37-dgA dimer and trimer (Messmann et al., 2000). It is not clear if this aggregation of HD37-dgA was related to enhanced toxicity. Clinical response in this trial was moderate including two PRs.

Blocked ricin ITs in NHL

The IT anti-B4-bR consists of blocked ricin and the murine MAb anti-B4, which binds to the CD19 cell-surface antigen (Nadler et al., 1983). Anti-B4-bR is the blocked ricin ITs analyzed most extensively in clinical trials (Table 4.2). Anti-B4-bR was effective in different SCID mice models against CD19-expressing cell lines (Shah et al., 1993a) and in cynomolgus monkeys (Shah et al., 1993b). Grossbard and coworkers subsequently performed the first phase I study in 25 patients with refractory or resistant B-cell neoplasms including NHL, chronic lymphocytic leukemia (CLL), and acute lymphoblastic leukemia (ALL) (Grossbard et al., 1992). Anti-B4-bR was administered daily as 1-h bolus infusion for 5 consecutive days with doses ranging from 1 to 60 µg/kg/day. Serum levels above 1 nmol/l were achieved transiently in most patients treated at the MTD of 50 µg/kg/day for 5 days. Compared with dgA-based ITs the DLT with blocked ricin-based ITs was different manifesting as transient, reversible grade III elevations in hepatic transaminases, without impaired hepatic synthetic function. Minor toxicities included transient hypoalbuminemia, thrombocytopenia, and fevers. Nine patients made HAMA and/or HARA within 1 month following therapy. One patient achieved a CR and two patients PRs.

In a phase II trial in 16 patients with relapsed CD19-positive NHL anti-B4-bR was administered at the previously established MTD (50 µg/kg/day) using a bolus infusion for 5 consecutive days (Multani et al., 1998). Toxicity was similar to what has been described previously. Six patients made HARA and three patients developed HAMA. No clinical responses were documented. The poor response might be explained by detection of anti-B4-bR in only one of seven tissue samples removed after IT infusion. In contrast, anti-B4-bR was detected in three of four bone marrow aspirates. Thus, anti-B4-bR did not consistently penetrate in all sites of disease. A similar distinct pattern of tissue penetration was observed with the humanized anti-CD52 MAb alemtuzumab (Osterborg et al., 1996; Osterborg et al., 1997).

To possibly overcome the major shortcoming of the bolus application of anti-B4-bR, namely the evidence of therapeutic serum levels only over a period of 4 to 6 h, a new phase I trial with continuous infusion over 7 days was initiated (Grossbard et al., 1993b). Thirty-four patients with resistant B-cell malignancies (26 NHL, 4 CLL, 4 ALL) received 7-day continuous infusion of anti-B4-bR at doses of 10 to 70 µg/kg/day for 7 days. The initial three cohorts of patients (10, 20, and 30 µg/kg/day $\times$ 7 days) also received a bolus infusion of 20 µg/kg before start of the continuous infusion. The MTD was reached at 50 µg/kg/day. Compared to bolus application, toxicity was higher with DLT represented by reversible grade IV elevations of hepatic transaminases and grade IV thrombocytopenia. VLS, myalgia, nausea, and fever were also reported. The pharmacokinetic profile demonstrated therapeutic serum levels after 48 h lasting over the 7 day period. Lower serum levels were achieved in patients with CLL and ALL and high numbers of peripheral tumor cells. Of 34 patients developed HAMA and/or HARA. CRs were reported in two patients presenting with low tumor mass and a PR in further three patients. In summary, patients treated with continuous infusions achieved a higher MTD accompanied by more severe side effects including VLS, nausea, fever, and myalgia. This might be due to a prolonged exposition of the endothelial cells to the IT resulting in nonspecific uptake.

Table 4.2 Clinical trials with blocked ricin A-chain ITs in NHL patients

Disease	Antigen	Construct	Application	Toxicity (MTD)	Immunogenicity	Response	Reference
B-NHL, CLL, ALL	CD19	Anti-B4-bR	1-h infusion Day 1–5	Transaminase elevation, VLS, thrombocytopenia (250 μg/kg)	9/25 HAMA/HARA	1/25 CR 2/25 PR	(Grossbard et al., 1992)
B-NHL, CLL, ALL	CD19	Anti-B4-bR	1-h infusion Day 1–7	Transaminase elevation, VLS, thrombocytopenia (350 μg/kg)	19/34 HAMA 18/34 HARA	2/34 CR 3/34 PR	(Grossbard et al., 1993b)
B-NHL	CD19	Anti-B4-bR	1-h infusion Day 1–5	Transaminase elevation, VLS, thrombocytopenia	3/16 HAMA 6/16 HARA	none	(Multani et al., 1998)
B-NHL post-ABMT	CD19	Anti-B4-bR	Continuous infusion Day1–7	Transaminase elevation, VLS, thrombocytopenia (280 μg/kg)	5/12 HAMA 7/12 Hara	7/12 patients CCR 31–67 months post-ABMT	(Grossbard et al., 1993a)
B-NHL post-ABMT	CD19	Anti-B4-bR	Continuous infusion Day 1–7	Transaminase elevation, VLS, thrombocytopenia	2/49 HAMA 23/49 HARA	27/49 patients in CCR in median 37 months post-ABMT	(Grossbard et al., 1999)
B-NHL post-CX	CD19	Anti-B4-bR	Continuous infusion Day 1–7	Transaminase elevation, VLS, thrombocytopenia	27/35 HARA	11/44 patients in CCR at 5 years	(Longo et al., 2000)
B-NHL HIV-related	CD19	Anti-B4-bR	Continuous infusion Day 1–28	Transaminase elevation, VLS, thrombocytopenia	1/9 HAMA 3/9 HARA	1/9 CR 1/9 PR	(Tulpule et al., 1994)
B-NHL HIV-related with CX	CD19	Anti-B4-bR	Continuous infusion Day 1–7 (140 μg/kg)	Transaminase elevation, flue-like symptoms	4/26 HAMA 8/26 HARA	13/26 CR 12/26 PR	(Scadden et al., 1998)

Abbreviations: ALL, acute lymphoblastic leukemia; NHL, non-Hodgkin's lymphoma; CLL, chronic lymphocytic leukemia; ABMT, autologous bone marrow transplantation; CX, chemotherapy; VLS, vascular leak syndrome; HAMA, human anti-mouse antibodies; HARA, human anti-ricin antibodies; CR, complete remission; CCR, continuous CR; PR, partial remission.

This observation is contrary to results from preclinical animal studies (O'Toole et al., 1998; Shah et al., 1993b).

In another phase I trial, anti-B4-bR was administered in an adjuvant setting to patients with chemosensitive relapsed B-NHL in CR after ABMT (Grossbard et al., 1993a). Twelve patients were treated with a 7 day continuous infusion at 20, 40, and 50 μg/kg/day. Retreatment was performed every 28 days. Potentially therapeutic serum levels could be sustained for 3 to 4 days. The MTD was 40 μg/kg/day for 7 days, which was lower compared with prior trials due to minimal or negligible tumor burden. The DLTs were reversible grade IV thrombocytopenia and elevation of hepatic transaminases. The comparatively mild VLS observed was not dose-limiting. Despite imunosuppression after ABMT, five patients made HAMA and seven patients made HARA. Eleven patients remained in CR between 13 and 26 months post-ABMT (median 17 months). In an updated report seven patients remained in CR within a median follow-up of more than 4 years (O'Toole et al., 1998). In a subsequent phase II study, anti-B4-bR was administered at a median of 112 days post-ABMT at a reduced dose of 30 μg/kg/day over 7 days every 14 days in 49 patients with B-cell NHL in CR (Grossbard et al., 1999). Since anti-B4-bR is a lipophobic compound dosing was modified to reflect the lean body mass (LBM) rather than the actual weight. The mean serum level on day 7 of the first course was 0.77+/−0.41 nM. Toxicity was clearly reduced compared with the prior phase I trial. Reversible toxicities included hepatic transaminase elevations, thrombocytopenia, myalgia, fatigue, nausea, and VLS. Twenty-three patients developed HAMA and/or HARA at a median of 22 days from the initiation of treatment. The 4 year disease-free survival and overall survival were estimated at 56% and 72% indicating continued relapse, respectively. Twenty-six patients remained in CR after a median follow-up of 54.5 months. This study demonstrated that anti-B4-bR can be administered safely to patients in an adjuvant setting. Based on these results, the Cancer and Acute Leukemia Group B (CALGB) initiated a randomized multicenter phase III study in which patients in CR received either anti-B4-bR after HDCT 60 and 120 days after ABMT or were observed without any additive treatment.

A similar phase II trial was initiated in advanced-stage indolent B-NHL patients who were in CR or PR after conventional polychemotherapy (Longo et al., 2000). Forty-four patients with minimal residual disease received six cycles of ProMACE-CytaBOM (prednisone, methotrexate, doxorubicin, cyclophosphamide, etoposide-cytarabine, bleomycin, vincristine, mechlorethamine) followed by a 7 day continuous infusion of anti-B4-blocked ricin IT at 30 μg/kg/day given every 14 days for up to six cycles. A median number of two courses was delivered due to the development of HARA (27 of 35 patients). Toxicity was very similar to prior trials. CR was achieved in 25 of 44 patients (57%), 22 after chemotherapy. Three patients converted from PR to CR after IT application. Interestingly, patients with PCR positivity of bcl-2 after all treatment relapsed, whereas the patients converting from bcl-2 positivity to bcl-2 negativity after IT administration remained in CR. Median duration of remission was 2 years. With a median follow-up of 5 years, 14 of 25 patients with CR have relapsed (56%) and overall survival was 61%. In conclusion, polychemotherapy in combination with anti-B4-bR did not produce durable CRs in the majority of patients with indolent lymphoma.

Other clinical investigations explored anti-B4-bR in HIV-related NHL in combination with polychemotherapy (Scadden et al., 1998) and to purge harvested bone marrow cells of patients with NHL before reinfusion of the bone marrow cells (Roy et al., 1995).

Ricin A-chain ITs in HL

A vast amount of murine MAbs against CD25 and CD30 was tested for their potential use against Hodgkin-derived cell lines without identifying a MAb with reasonable intrinsic cytotoxicity (Engert et al., 1990; Engert et al., 1991). Thus, ricin A-chain ITs against Hodgkin-associated antigens were developed.

Ricin A-chain ITs against CD25

A potential target for a selective immunotherapy of HL is the IL-2 receptor, which is expressed in high amounts on the vast majority of H-RS cells (Strauchen and Breakstone, 1987). The IL-2 receptor is composed of three different membrane components (α-, β-, γ-chain) (Taniguchi and Minami, 1993). CD25, a 55 kD glycoprotein, represents the α-chain and is undetectable in resting lymphocytes and stem cells but is efficiently induced upon T-cell activation. The most potent anti-CD25 IT, RFT5.dgA, inhibited the protein synthesis of H-RS cells (L540Cy) at a concentration (IC_{50}) of 7×10^{-12} M, which is nearly identical to that of native ricin under the same experimental conditions (Engert et al., 1991). RFT5.dgA showed no major cross-reactivity with any human tissue other than lymphoid, where a few large cells in tonsils and lymph nodes were stained. This construct was evaluated in triple-beige nude mice with subcutaneous Hodgkin tumors and SCID mice with disseminated human Hodgkin tumors inducing CRs in 78% and 95%, respectively (Engert et al., 1991; Winkler et al., 1994).

RFT5.dgA was subsequently selected for a phase I/II clinical trial in 15 patients with refractory HD (Engert et al., 1997). The IT was administered i.v. over 4 h every other day for 7 days. The MTD was determined at 15 mg/m². Side effects were related to transient VLS. Two patients experienced an allergic reaction with generalized urticaria and mild bronchospasm. Seven of fifteen patients made HARA >1 μg/ml and six patients developed HAMA >1 μg/ml. Maximum serum concentrations ranged from 0.2 to 9.7 μg/ml with a half-life of 4.0 to 10.5 h. Subsequently, further patients were treated at MTD to better assess the efficacy of RFT5.dgA in heavily pretreated HL patients (Schnell et al., 2000). Therapy was well tolerated with grade 3 toxicities (VLS, myalgia) in five of eighteen patients. Response of seventeen evaluable patients included two PRs, one minor response (MR), and five SD. One reason for the moderate response might be the strong expression of CD25 on only 30% of H-RS cells.

Ricin A-chain ITs against CD30

The lymphocyte activation marker CD30 is expressed consistently on the malignant cells in HL and has been shown to be an excellent potential target for immunotherapy of human HL. The CD30 antigen was originally discovered on cultured H-RS cells using the MAb Ki-1 (Stein et al., 1985). Normal human organs revealed no major cross-reactivity of CD30 MAbs. The most effective ricin A-chain IT, Ki-4.dgA, was five times more potent *in vitro* compared to former CD30 A-chain ITs and demonstrated high efficacy in the treatment of disseminated human HL in a SCID mouse model (Schnell et al., 1995). Thus, Ki-4.dgA was selected for a clinical phase I trial in 17 heavily pretreated patients with refractory CD30 positive HL and NHL (Schnell et al., 2002). The IT was given in four bolus infusions every other day in escalating doses. Peak serum concentrations of the intact IT varied from 0.23 to 1.7 μg/ml. Side effects were related to the VLS and the MTD was reached at 5 mg/m^2. The low MTD could be in part explained by binding of the intact IT to soluble CD30 after infusion of the IT. Forty percent of patients made HARA and one patient made HAMA. Responses included one PR and one MR. This lower than expected response rate might be at least in part related to the low dose of IT administered.

In conclusion, RFT5.dgA and Ki-4.dgA given to patients with resistant HL demonstrated moderate tolerability and efficacy (Table 4.3). One of the explanations for these results were the unfavorable group of patients selected in these studies. Most patients were heavily pretreated with multiple prior chemotherapies presenting with highly active disease and large tumor burden. VLS occurred at lower doses in the Ki-4.dgA trial compared with the RFT5.dgA trial and similar trials with other ITs in patients with NHL. This might be due to the small numbers of CD30$^+$ peripheral blood mononuclear cells (PBMCs), the binding of Ki-4.dgA to sCD30, and the lack of shed antigens such as CD19 and CD22 in NHL patients. A strong inverse correlation between circulating tumor cells and toxicity has been reported in other trials (Messmann et al., 2000). In the RFT5.dgA trial, binding of RFT5.dgA to sCD25 was not detected even though it bound to CD25$^+$ PBMCs. To reduce binding of Ki-4.dgA to sCD30 it might be prudent to infuse the native MAb before treatment with the IT. Since metalloproteinases induce the shedding of CD30, the blockade by hydroxamic acid-based metalloproteinase inhibitors might also reduce toxicity (Hansen et al., 1995; Hansen et al., 2002). In contrast to the trials in NHL patients, in HL patients neither pulmonary edema nor aphasia related to VLS was observed (Schnell et al., 2003). Since > 50% of the patients in the HL trials had pulmonary HL other reasons must be considered. As well in contrast to the NHL studies nearly all HL patients received irradiation therapy before IT treatment without influencing VLS. The most obvious explanation apart from the different histology is the younger age of patients enrolled in the HL trials (30 to 35 years) compared with 49 to 60 years in the NHL trials. Thus, different symptoms of VLS might be prevalent in patients of different ages or suggesting that HL and NHL have different predisposing

Table 4.3 Clinical trials with ricin A-chain ITs in HL patients

Disease	Antigen	Construct	Application	Toxicity (MTD)	Immunogenicity	Response	Reference
HL	CD25	RFT5.dgA	4-h infusion, days 1–3–5–7	VLS, myalgia, nausea, fatigue (15 mg/m^2)	6/15 HAMA 7/15 HARA	2/15 PR	(Engert et al., 1997)
HL	CD25	RFT5.dgA	4-h infusion, days 1–3–5–7	VLS, myalgia, nausea, fatigue (15 mg/m^2)	11/18 HAMA 11/18 HARA	2/17 PR	(Schnell et al., 2000)
HL	CD30	Ki-4.dgA	4-h infusion, days 1–3–5–7	VLS, myalgia, nausea, fatigue (5 mg/m^2)	1/17 HAMA 1/17 HARA	1/17 PR	(Schnell et al., 2002)

Abbreviations: HL, Hodgkin's lymphoma; VLS, vascular leak syndrome; HAMA, human anti-mouse antibodies; HARA, human anti-ricin; PR, partial remission.

factors that have not been identified yet. The moderate response rate might be explained by the relatively low dose of administered IT in the anti-CD30 IT trial. In the anti-CD25 IT trial this might be related to a strong expression of CD25 on only 30% of H-RS cells. Thus, the IL-2 receptor or the CD25 antigen might not be the optimal target in HL, which is supported by other trials with recombinant anti-CD25 ITs or anti-IL-2 constructs demonstrating very similar disappointing response rates (Kreitman et al., 2000; LeMaistre et al., 1992; LeMaistre et al., 1993; LeMaistre et al., 1998).

Obstacles of ricin ITs

Clinical trials with A-chain ITs or blocked ricin ITs in patients with relapsed or refractory B-cell NHL or HL demonstrated efficacy but also revealed obstacles. ITs containing deglycosylated ricin A-chain hepatotoxicity was successfully eliminated. This was however the DLT in trials using blocked ricin. Some patients in the blocked ricin trials developed reversible thrombocytopenia but a more pronounced myelosuppression was not reported. The most common side effect was VLS, which was described in all trials and defined the DLT in the dgA trials. A further possible obstacle to a prolonged clinical application in either ricin conjugate was the development of antibodies against the two components of the IT.

Vascular leak syndrome

VLS is clinically characterized by hypoalbuminemia with consecutive edema, weight gain, hypotension, dyspnea, tachycardia, and myalgia in patients without intrinsic cardiac, renal, or hepatic disease. Severe VLS caused some deaths in trials with dgA ITs. A retrospective analysis of 102 NHL patients treated with two different dgA ITs (RFB4.dgA, HD37.dgA) indicates that the VLS was more frequent and more severe in patients who had prior irradiation (Schindler et al., 2001) leading to the suggested consequence to exclude irradiated patients from IT trials. Other clinical predictors for VLS are the absence of even very small numbers of circulating tumor cells and sustained IT serum levels >1 µg/ml (Messmann et al., 2000). The VLS had previously been observed in patients treated with IL-2 (Rosenstein et al., 1986). Thus, initial speculations of VLS in IT treatment focused on increases of cytokines. Efforts to document a potential involvement of a variety of different cytokines including IL-1, IL-2, IL-4, IL-6, IL-10, or soluble adhesion molecules like sELAM or sICAM during IT treatment have failed (Sausville et al., 1995). In patients with severe VLS higher levels of tumor necrosis factor (TNF-α) were observed (Baluna et al., 1996b). *In vitro* models (human umbilical vein endothelial cells, HUVEC) studying VLS indicate a dramatic change in morphology of HUVECs after treatment with ricin A-chain ITs (Soler-Rodriguez et al., 1993). In addition, the permeability of HUVEC increased related to morphologic changes, which appeared just 1 h after exposure to the toxins, whereas inhibition of protein synthesis was not detectable until 4 h after exposure. This effect of dgA ITs on HUVECs can be suppressed in the presence of fibronectin (Fn), an extracellular matrix protein, which plays a role in the maintenance of vascular integrity (Baluna et al., 1996a), suggesting an interference with Fn-mediated adhesion. Recently, it was demonstrated that a three amino acid sequence motif, (x)D(y), where x are L, I, G, or V and y are V, L, or S, in toxins and IL-2 damages endothelial cells (Baluna et al., 1999). Thus, when peptides from ricin A-chain containing this sequence motif are linked to murine IgG, they bind to and damage endothelial cell both *in vitro* and *in vivo*. A detailed analysis of the three amino acid sequence LDV demonstrated binding to and damage of endothelial cells and the induction of early manifestations of apoptosis in HUVECs by activating caspase-3 (Baluna et al., 2000). These data might indicate that ricin A-chain-mediated inhibition of protein synthesis (due to its active site) and apoptosis (due to LDV) are mediated by different sites of the toxin. Thus, mutations or deletions of LDV may obtain the ability to eradicate tumor cells in the absence of endothelial cell-mediated VLS (Smallshaw et al., 2003).

Immunogenicity

The development of HAMA and HARA has been reported in most clinical phase I/II trials. These antibodies can neutralize circulating IT in the peripheral blood of patients resulting in decreased half-life and ineffective killing of tumor cells. Attempts to reduce the immune response by coadministration of immunosuppressive agents have been unsuccessful so far. The development of chimeric, humanized, or human ligands (Morrison et al., 1984; Woodard et al., 1998; Lonberg et al., 1994; Huhn et al., 2001; Zewe et al., 1997) and toxins (Huhn et al., 2001; Zewe et al., 1997) might solve this problem. The extensive experience with the chimeric anti-CD20 antibody rituximab demonstrated human anti-chimeric antibodies (HACA) in less than 1% (Maeda et al., 2001; McLaughlin et al., 1998) compared to HAMA response in up to 60% in trials using ricin-based constructs.

Summary and perspectives

Several clinical trials in B-NHL and HL using ricin-based ITs have been conducted from which the following conclusions emerge.

(1) The different ITs demonstrated biologic activity, even in heavily pretreated patients with bulky relapsed disease. Response rates varied from 6% to 23%. (2) VLS, myalgia, transient liver enzyme elevations, and thrombocytopenia are the most common side effects. It seems that the "therapeutic window," which is needed between concentrations mediating targeted efficacy and nontargeted toxicity is relatively small. (3) Immunogenicity of ricin-based ITs compromises cytotoxicity and the potential to deliver repeat courses of therapy. (4) No randomized phase III trial has been published yet, which demonstrated a benefit to administer ITs.

Since ricin-based ITs were developed before the introduction of highly active MAbs like the chimeric MAb rituximab (Maloney et al., 1997; Davis et al., 1999; Coiffier et al., 2002; McLaughlin et al., 1998) or the humanized MAb alemtuzumab (Osterborg et al., 1997), it is questionable if ITs can compete with these molecules. Rituximab combines effectiveness and low toxicity, which is infusion related. Other promising constructs in B-NHL are radioimmunoconstructs such as yttrium-90 ibritumomab tiuxetan and iodine-131 tositumomab, which also demonstrated impressive response rates with tolerable toxicity (Witzig et al., 1999; Witzig et al., 2002; Kaminski et al., 2001). These two constructs are commercially available for refractory B-NHL. Recently, chimeric or complete human anti-CD30 MAbs were described and are currently under clinical investigation in HL (Wahl et al., 2002; Barlett, 2002; Borchmann et al., 2003). Immunoconjugates like bispecific MAbs showed impressive antitumor response in refractory HL patients without reaching the MTD (Borchmann et al., 2002; Hartmann et al., 2001; Hartmann et al., 1997). Thus, the clinically tested ricin-based ITs might be of moderate clinical value due to the small therapeutic window. Genetic engineered ITs with mutations in the three-amino acid motif inducing VLS might solve this problem. The newly developed mutant RFB4-N97A was more effective in xenografted SCID mice compared to ricin A-chain used before without causing VLS at the same dose (Smallshaw et al., 2003). Further clinical trials with this new generation of ricin-based ITs are needed to define a potential role in lymphoma therapy.

References

Agnarsson, B.A. and Kadin, M.E. (1989) *Cancer*, 63, 2083–7.

Amlot, P.L., Stone, M.J., Cunningham, D., Fay, J., Newman, J., Collins, R., May, R., McCarthy, M., Richardson, J., Ghetie, V, et al. (1993) *Blood*, 82, 2624–33.

Anderson, K.C., Bates, M.P., Slaughenhoupt, B.L., Pinkus, G.S., Schlossman, S.F., and Nadler, L.M. (1984) *Blood*, 63, 1424–33.

Baker, D.G. and Krochak, R.J. (1989) *Cancer Invest*, 7, 287–94.

Baluna, R., Coleman, E., Jones, C., Ghetie, V., and Vitetta, E. S. (2000) *Exp Cell Res*, 258, 417–24.

Baluna, R., Ghetie, V., Oppenheimer-Marks, N., and Vitetta, E. S. (1996a) *Int J Immunopharmacol*, 18, 355–61.

Baluna, R., Rizo, J., Gordon, B.E., Ghetie, V., and Vitetta, E.S. (1999) *Proc Natl Acad Sci USA*, 96, 3957–62.

Baluna, R., Sausville, E.A., Stone, M.J., Stetler-Stevenson, M.A., Uhr, J.W., and Vitetta, E.S. (1996b) *Clin Cancer Res*, 2, 1705–12.

Banchereau, J. and Rousset, F. (1992) *Adv Immunol*, 52, 125–262.

Barlett, N.L., Younes, A., Carabasi, M.A., Espina, B., DiPersio, J.F., Schliebner, S.D., Siegall, C., and Sing, A.P. (2002) *Blood*, 100.

Bertomeu, M.C., Gallo, S., Lauri, D., Levine, M.N., Orr, F.W., and Buchanan, M.R. (1990) *Clin Exp Metastasis*, 8, 511–8.

Blakey, D.C., Watson, G.J., Knowles, P.P., and Thorpe, P.E. (1987) *Cancer Res*, 47, 947–52.

Borchmann, P., Schnell, R., Fuss, I., Manzke, O., Davis, T., Lewis, L.D., Behnke, D., Wickenhauser, C., Schiller, P., Diehl, V., and Engert, A. (2002) *Blood*, 100, 3101–7.

Borchmann, P., Treml, J.F., Hansen, H., Gottstein, C., Schnell, R., Staak, O., Zhang, H.F., Davis, T., Keler, T., Diehl, V., Graziano, R.F., and Engert, A. (2003) *Blood*, 102(10), 3737–42.

Byers, V.S., Henslee, P.J., Kernan, N.A., Blazar, B.R., Gingrich, R., Phillips, G.L., LeMaistre, C.F., Gilliland, G., Antin, J.H., Martin, P., et al. (1990) *Blood*, 75, 1426–32.

Coiffier, B., Lepage, E., Briere, J., Herbrecht, R., Tilly, H., Bouabdallah, R., Morel, P., Van Den Neste, E., Salles, G., Gaulard, P., Reyes, F., Lederlin, P., and Gisselbrecht, C. (2002) *N Engl J Med*, 346, 235–42.

Conry, R.M., Khazaeli, M.B., Saleh, M.N., Ghetie, V., Vitetta, E.S., Liu, T., and LoBuglio, A.F. (1995) *J Immunother Emphasis Tumor Immunol*, 18, 231–41.

Davis, T.A., White, C.A., Grillo-Lopez, A.J., Velasquez, W.S., Link, B., Maloney, D.G., Dillman, R.O., Williams, M.E., Mohrbacher, A., Weaver, R., Dowden, S., and Levy, R. (1999) *J Clin Oncol*, 17, 1851–7.

Endo, Y., Mitsui, K., Motizuki, M., and Tsurugi, K. (1987) *J Biol Chem*, 262, 5908–12.

Endo, Y. and Tsurugi, K. (1987) *J Biol Chem*, 262, 8128–30.

Endo, Y. and Tsurugi, K. (1988) *Nucleic Acids Symp Ser*, 19, 139–42.

Engert, A., Burrows, F., Jung, W., Tazzari, P.L., Stein, H., Pfreundschuh, M., Diehl, V., and Thorpe, P. (1990) *Cancer Res*, 50, 84–8.

Engert, A., Diehl, V., Schnell, R., Radszuhn, A., Hatwig, M.T., Drillich, S., Schon, G., Bohlen, H., Tesch, H., Hansmann, M.L., Barth, S., Schindler, J., Ghetie, V., Uhr, J., and Vitetta, E. (1997) *Blood*, 89, 403–10.

Engert, A., Gottstein, C., Bohlen, H., Winkler, U., Schon, G., Manske, O., Schnell, R., Diehl, V., and Thorpe, P. (1995) *Int J Cancer*, 63, 304–9.

Engert, A., Martin, G., Amlot, P., Wijdenes, J., Diehl, V., and Thorpe, P. (1991) *Int J Cancer*, 49, 450–6.

Engert, A., Sausville, E.A., and Vitetta, E. (1998) *Curr Top Microbiol Immunol*, 234, 13–33.

Fulton, R.J., Uhr, J.W., and Vitetta, E.S. (1988) *Cancer Res*, 48, 2626–31.

Ghetie, M.A., Podar, E.M., Gordon, B.E., Pantazis, P., Uhr, J.W., and Vitetta, E. S. (1996) *Int J Cancer*, 68, 93–6.

Ghetie, M.A., Tucker, K., Richardson, J., Uhr, J.W., and Vitetta, E.S. (1992) *Blood*, 80, 2315–20.

Gribben, J.G., Freedman, A.S., Neuberg, D., Roy, D.C., Blake, K.W., Woo, S.D., Grossbard, M.L., Rabinowe, S.N., Coral, F., Freeman, G.J., et al. (1991) *N Engl J Med*, 325, 1525–33.

Gribben, J.G., Neuberg, D., Freedman, A.S., Gimmi, C.D., Pesek, K.W., Barber, M., Saporito, L., Woo, S. D., Coral, F., Spector, N., et al. (1993) *Blood*, 81, 3449–57.

Grossbard, M.L., Freedman, A.S., Ritz, J., Coral, F., Goldmacher, V.S., Eliseo, L., Spector, N., Dear, K., Lambert, J.M., Blattler, W.A., et al. (1992) *Blood*, 79, 576–85.

Grossbard, M.L., Gribben, J.G., Freedman, A.S., Lambert, J.M., Kinsella, J., Rabinowe, S.N., Eliseo, L., Taylor, J.A., Blattler, W.A., Epstein, C.L., et al. (1993a) *Blood*, 81, 2263–71.

Grossbard, M.L., Lambert, J.M., Goldmacher, V.S., Spector, N.L., Kinsella, J., Eliseo, L., Coral, F., Taylor, J.A., Blattler, W.A., Epstein, C.L., et al. (1993b) *J Clin Oncol*, 11, 726–37.

Grossbard, M.L., Multani, P.S., Freedman, A.S., O'Day, S., Gribben, J.G., Rhuda, C., Neuberg, D., and Nadler, L.M. (1999) *Clin Cancer Res*, 5, 2392–8.

Hansen, H.P., Kisseleva, T., Kobarg, J., Horn-Lohrens, O., Havsteen, B., and Lemke, H. (1995) *Int J Cancer*, 63, 750–6.

Hansen, H.P., Matthey, B., Barth, S., Kisseleva, T., Mokros, T., Davies, S.J., Beckett, R.P., Foelster-Holst, R., Lange, H.H., Engert, A., and Lemke, H. (2002) *Int J Cancer*, 98, 210–5.

Hartmann, F., Renner, C., Jung, W., da Costa, L., Tembrink, S., Held, G., Sek, A., Konig, J., Bauer, S., Kloft, M., and Pfreundschuh, M. (2001) *Clin Cancer Res*, 7, 1873–81.

Hartmann, F., Renner, C., Jung, W., Deisting, C., Juwana, M., Eichentopf, B., Kloft, M., and Pfreundschuh, M. (1997) *Blood*, 89, 2042–7.

Holden, C.A., Staughton, R.C., Campbell, M.A., and MacDonald, D.M. (1982) *J Am Acad Dermatol*, 6, 507–13.

Huhn, M., Sasse, S., Tur, M.K., Matthey, B., Schinkothe, T., Rybak, S.M., Barth, S., and Engert, A. (2001) *Cancer Res*, 61, 8737–42.

Kaminski, M.S., Zelenetz, A.D., Press, O.W., Saleh, M., Leonard, J., Fehrenbacher, L., Lister, T.A., Stagg, R.J., Tidmarsh, G.F., Kroll, S., Wahl, R.L., Knox, S.J., and Vose, J.M. (2001) *J Clin Oncol*, 19, 3918–28.

Kanzler, H., Hansmann, M.L., Kapp, U., Wolf, J., Diehl, V., Rajewsky, K., and Kuppers, R. (1996) *Blood*, 87, 3429–36.

Knowles, P.P. and Thorpe, P.E. (1987) *Anal Biochem*, 160, 440–3.

Kreitman, R.J., Wilson, W.H., White, J.D., Stetler-Stevenson, M., Jaffe, E.S., Giardina, S., Waldmann, T.A., and Pastan, I. (2000) *J Clin Oncol*, 18, 1622–36.

Krolick, K.A., Uhr, J.W., Slavin, S., and Vitetta, E.S. (1982) *J Exp Med*, 155, 1797–809.

Kuan, C.T., Pai, L.H., and Pastan, I. (1995) *Clin Cancer Res*, 1, 1589–94.

Lambert, J.M., Goldmacher, V.S., Collinson, A.R., Nadler, L.M., and Blattler, W.A. (1991a) *Cancer Res*, 51, 6236–42.

Lambert, J.M., McIntyre, G., Gauthier, M.N., Zullo, D., Rao, V., Steeves, R.M., Goldmacher, V.S., and Blattler, W.A. (1991b) *Biochemistry*, 30, 3234–47.

LeMaistre, C.F., Craig, F.E., Meneghetti, C., McMullin, B., Parker, K., Reuben, J., Boldt, D.H., Rosenblum, M., and Woodworth, T. (1993) *Cancer Res*, 53, 3930–4.

LeMaistre, C.F., Meneghetti, C., Rosenblum, M., Reuben, J., Parker, K., Shaw, J., Deisseroth, A., Woodworth, T., and Parkinson, D.R. (1992) *Blood*, 79, 2547–54.

LeMaistre, C.F., Rosen, S., Frankel, A., Kornfeld, S., Saria, E., Meneghetti, C., Drajesk, J., Fishwild, D., Scannon, P., and Byers, V. (1991) *Blood*, 78, 1173–82.

LeMaistre, C.F., Saleh, M.N., Kuzel, T.M., Foss, F., Platanias, L. C., Schwartz, G., Ratain, M., Rook, A., Freytes, C. O., Craig, F., Reuben, J., and Nichols, J. C. (1998) *Blood*, 91, 399–405.

Leonard, J.P. and Link, B.K. (2002) *Semin Oncol*, 29, 81–6.

Lindstrom, A.L., Erlandsen, S.L., Kersey, J.H., and Pennell, C.A. (1997) *Blood*, 90, 2323–34.

Lonberg, N., Taylor, L.D., Harding, F.A., Trounstine, M., Higgins, K.M., Schramm, S.R., Kuo, C.C., Mashayekh, R., Wymore, K., McCabe, J.G., et al. (1994) *Nature*, 368, 856–9.

Longo, D.L., Duffey, P.L., Gribben, J.G., Jaffe, E.S., Curti, B.D., Gause, B.L., Janik, J.E., Braman, V.M., Esseltine, D., Wilson, W.H., Kaufman, D., Wittes, R.E., Nadler, L.M., and Urba, W.J. (2000) *Cancer J*, 6, 146–50.

Maeda, T., Yamada, Y., Tawara, M., Yamasaki, R., Yakata, Y., Tsutsumi, C., Onimaru, Y., Kamihira, S., and Tomonaga, M. (2001) *Int J Hematol*, **74**, 70–5.

Maloney, D.G. (1999) *Semin Oncol*, 26, 74–8.

Maloney, D.G., Grillo-Lopez, A.J., White, C.A., Bodkin, D., Schilder, R.J., Neidhart, J.A., Janakiraman, N., Foon, K.A., Liles, T.M., Dallaire, B.K., Wey, K., Royston, I., Davis, T., and Levy, R. (1997) *Blood*, 90, 2188–95.

McLaughlin, P., Grillo-Lopez, A.J., Link, B.K., Levy, R., Czuczman, M.S., Williams, M.E., Heyman, M.R., Bence-Bruckler, I., White, C.A., Cabanillas, F., Jain, V., Ho, A.D., Lister, J., Wey, K., Shen, D., and Dallaire, B.K. (1998) *J Clin Oncol*, 16, 2825–33.

Messmann, R.A., Vitetta, E.S., Headlee, D., Senderowicz, A.M., Figg, W.D., Schindler, J., Michiel, D.F., Creekmore, S., Steinberg, S.M., Kohler, D., Jaffe, E.S., Stetler-Stevenson, M., Chen, H., Ghetie, V., and Sausville, E.A. (2000) *Clin Cancer Res*, 6, 1302–13.

Morrison, S.L., Johnson, M.J., Herzenberg, L.A., and Oi, V.T. (1984) *Proc Natl Acad Sci USA*, 81, 6851–5.

Multani, P.S., O'Day, S., Nadler, L.M., and Grossbard, M.L. (1998) *Clin Cancer Res*, 4, 2599–604.

Nadler, L.M., Anderson, K.C., Marti, G., Bates, M., Park, E., Daley, J.F., and Schlossman, S.F. (1983), *J Immunol*, 131, 244–50.

Osterborg, A., Dyer, M.J., Bunjes, D., Pangalis, G.A., Bastion, Y., Catovsky, D., and Mellstedt, H. (1997) *J Clin Oncol*, 15, 1567–74.

Osterborg, A., Fassas, A.S., Anagnostopoulos, A., Dyer, M.J., Catovsky, D., and Mellstedt, H. (1996) *Br J Haematol*, 93, 151–3.

O'Toole, J.E., Esseltine, D., Lynch, T.J., Lambert, J.M., and Grossbard, M.L. (1998) *Curr Top Microbiol Immunol*, 234, 35–56.

Riethmuller, G., Schneider-Gadicke, E., Schlimok, G., Schmiegel, W., Raab, R., Hoffken, K., Gruber, R., Pichlmaier, H., Hirche, H., Pichlmayr, R., et al. (1994) *Lancet*, 343, 1177–83.

Rosenstein, M., Ettinghausen, S.E., and Rosenberg, S.A. (1986), *J Immunol*, 137, 1735–42.

Roy, D.C., Griffin, J.D., Belvin, M., Blattler, W.A., Lambert, J.M., and Ritz, J. (1991) *Blood*, 77, 2404–12.

Roy, D.C., Perreault, C., Belanger, R., Gyger, M., Le Houillier, C., Blattler, W.A., Lambert, J.M., and Ritz, J. (1995) *J Clin Immunol*, 15, 51–7.

Sausville, E.A., Headlee, D., Stetler-Stevenson, M., Jaffe, E.S., Solomon, D., Figg, W.D., Herdt, J., Kopp, W.C., Rager, H., Steinberg, S.M., et al. (1995) *Blood*, 85, 3457–65.

Scadden, D.T., Schenkein, D.P., Bernstein, Z., Luskey, B., Doweiko, J., Tulpule, A., and Levine, A.M. (1998) *Cancer*, 83, 2580–7.

Schindler, J., Sausville, E., Messmann, R., Uhr, J.W., and Vitetta, E.S. (2001) *Clin Cancer Res*, 7, 255–8.

Schnell, R., Borchmann, P., Staak, J.O., Schindler, J., Ghetie, V., Vitetta, E.S., and Engert, A. (2003) *Ann Oncol*, 14, 729–36.

Schnell, R., Linnartz, C., Katouzi, A.A., Schon, G., Bohlen, H., Horn-Lohrens, O., Parwaresch, R.M., Lange, H., Diehl, V., Lemke, H., et al. (1995) *Int J Cancer*, 63, 238–44.

Schnell, R., Staak, O., Borchmann, P., Schwartz, C., Matthey, B., Hansen, H., Schindler, J., Ghetie, V., Vitetta, E.S., Diehl, V., and Engert, A. (2002) *Clin Cancer Res*, 8, 1779–86.

Schnell, R., Vitetta, E., Schindler, J., Borchmann, P., Barth, S., Ghetie, V., Hell, K., Drillich, S., Diehl, V., and Engert, A. (2000) *Leukemia*, 14, 129–35.

Shah, S.A., Halloran, P.M., Ferris, C.A., Levine, B.A., Bourret, L. A., Goldmacher, V.S., and Blattler, W.A. (1993a) *Cancer Res*, 53, 1360–7.

Shah, S.A., Lambert, J.M., Goldmacher, V.S., Esber, H.J., Levin, J.L., Chungi, V., Zutshi, A., Braman, G.M., Ariniello, P.D., Taylor, J.A., et al. (1993b) *Int J Immunopharmacol*, 15, 723–36.

Simmons, B.M., Stahl, P.D., and Russell, J.H. (1986) *J Biol Chem*, 261, 7912–20.

Smallshaw, J.E., Ghetie, V., Rizo, J., Fulmer, J.R., Trahan, L.L., Ghetie, M.A., and Vitetta, E.S. (2003) *Nat Biotechnol*, 21, 387–91.

Soler-Rodriguez, A.M., Ghetie, M.A., Oppenheimer-Marks, N., Uhr, J.W., and Vitetta, E.S. (1993) *Exp Cell Res*, 206, 227–34.

Stein, H., Mason, D.Y., Gerdes, J., O'Connor, N., Wainscoat, J., Pallesen, G., Gatter, K., Falini, B., Delsol, G., Lemke, H., et al. (1985) *Blood*, 66, 848–58.

Stillmark, H. (1897) *J Exp Med*, 2, 197–216.

Stone, M.J., Sausville, E.A., Fay, J.W., Headlee, D., Collins, R.H., Figg, W.D., Stetler-Stevenson, M., Jain, V., Jaffe, E.S., Solomon, D., Lush, R.M., Senderowicz, A., Ghetie, V., Schindler, J., Uhr, J.W., and Vitetta, E.S. (1996) *Blood*, 88, 1188–97.

Strauchen, J.A. and Breakstone, B.A. (1987) *Am J Pathol*, 126, 506–12.

Taniguchi, T. and Minami, Y. (1993) *Cell*, 73, 5–8.

Thorpe, P.E., Detre, S.I., Foxwell, B.M., Brown, A.N., Skilleter, D.N., Wilson, G., Forrester, J.A., and Stirpe, F. (1985) *Eur J Biochem*, 147, 197–206.

Thorpe, P.E., Wallace, P.M., Knowles, P.P., Relf, M.G., Brown, A.N., Watson, G.J., Blakey, D.C., and Newell, D. R. (1988) *Cancer Res*, 48, 6396–403.

Thorpe, P.E., Wallace, P.M., Knowles, P.P., Relf, M.G., Brown, A.N., Watson, G.J., Knyba, R.E., Wawrzynczak, E.J., and Blakey, D.C. (1987) *Cancer Res*, 47, 5924–31.

Tulpule, A., Anderson, L., Levine, A.M., Espina, B., Esplin, J., Boswell, W., Scadden, D., Esseltine, D., and Epstein, C.L. (1994) *Proc Am Soc Clin Oncol*, 13, 52.

Vitetta, E.S., Cushley, W., and Uhr, J.W. (1983) *Proc Natl Acad Sci USA*, 80, 6332–5.

Vitetta, E.S., Fulton, R.J., May, R.D., Till, M., and Uhr, J.W. (1987) *Science*, 238, 1098–104.

Vitetta, E.S., Fulton, R.J., and Uhr, J.W. (1984) *J Exp Med*, 160, 341–6.

Vitetta, E.S., Stone, M., Amlot, P., Fay, J., May, R., Till, M., Newman, J., Clark, P., Collins, R., Cunningham, D., et al. (1991) *Cancer Res*, 51, 4052–8.

Vitetta, E.S. and Thorpe, P.E. (1991) *Semin Cell Biol*, 2, 47–58.

Vitetta, E.S., Thorpe, P.E., and Uhr, J.W. (1993) *Trends Pharmacol Sci*, 14, 148–54.

Wahl, A.F., Klussman, K., Thompson, J.D., Chen, J.H., Francisco, L.V., Risdon, G., Chace, D. F., Siegall, C.B., and Francisco, J.A. (2002) *Cancer Res*, 62, 3736–42.

Winkler, U., Gottstein, C., Schon, G., Kapp, U., Wolf, J., Hansmann, M.L., Bohlen, H., Thorpe, P., Diehl, V., and Engert, A. (1994) *Blood*, 83, 466–75.

Witzig, T.E., Gordon, L.I., Cabanillas, F., Czuczman, M.S., Emmanouilides, C., Joyce, R., Pohlman, B.L., Bartlett, N.L., Wiseman, G.A., Padre, N., Grillo-Lopez, A.J., Multani, P., and White, C.A. (2002) *J Clin Oncol*, 20, 2453–63.

Witzig, T.E., White, C.A., Wiseman, G.A., Gordon, L.I., Emmanouilides, C., Raubitschek, A., Janakiraman, N., Gutheil, J., Schilder, R.J., Spies, S., Silverman, D.H., Parker, E., and Grillo-Lopez, A.J. (1999) *J Clin Oncol*, 17, 3793–803.

Wolf, J., Kapp, U., Bohlen, H., Kornacker, M., Schoch, C., Stahl, B., Mucke, S., von Kalle, C., Fonatsch, C., Schaefer, H.E., Hansmann, M.L., and Diehl, V. (1996) *Blood*, 87, 3418–28.

Wood, G.S., Deneau, D.G., Miller, R.A., Levy, R., Hoppe, R.T., and Warnke, R.A. (1982) *Blood*, 59, 876–82.

Woodard, S.L., Fraser, S.A., Winkler, U., Jackson, D.S., Kam, C.M., Powers, J.C., and Hudig, D. (1998) *J Immunol*, 160, 4988–93.

Youle, R.J., Murray, G.J., and Neville, D.M., Jr. (1981) *Cell*, 23, 551–9.

Zewe, M., Rybak, S.M., Dubel, S., Coy, J.F., Welschof, M., Newton, D.L., a nd Little, M. (1997) *Immunotechnology*, 3, 127–36.

part three

Diphtheria toxin-based immunotoxins/cytotoxins

chapter five

Diphtheria-based immunotoxins: Targeting overexpressed interleukin-13 receptor and urokinase-type plasminogen activator receptor on glioblastoma multiforme

Edward Rustamzadeh, Walter A. Hall, Deborah Todhunter, and Daniel Vallera

Contents

Introduction

Malignant gliomas have a poor prognosis. The median survival for treated patients with glioblastoma multiforme (GBM) is less than 1 year with less than 20% surviving two years.[1] In addition to primary surgical intervention, adjuvant treatment with chemotherapy and radiation therapy has become the standard of care for treating brain tumors. Certain central nervous system (CNS) tumors are especially sensitive to chemotherapy such as oligoden-drogliomas and primary CNS lymphomas.[2-4] However, the majority of malignant CNS tumors do not respond to treatment.[5-7] There are numerous reasons why most treatment modalities have failed. Surgical resection will only treat to the tumor boundaries as defined by CT and MRI. These imaging techniques demonstrate the tumor boundary as the rim of contrast enhance-ment around the tumor where the blood-brain barrier (BBB) is disrupted. Unfortunately, blood-brain barrier disruption is greatest at the center of the tumor and less at the periphery;[8,9] therefore, it is not surprising that most tumor recurrences occur within 2 cm of the tumor resection margin.[10-12] Failure of chemotherapy regimens can be attributed to an inability to achieve therapeutic concentrations within the tumor because of the BBB (which prevents the passage of drugs with molecular weight greater than 180 kDa),[13] a low growth fraction for portions of tumors,[14] decreased uptake, and repair of drug-induced damage by the tumor.[15] Furthermore, the side effects of chemotherapy such as neutropenia, gastrointestinal ulceration, nephro-toxicity, neuropathy, and ototoxicity, limit the amount of drug that can be administered. Similarly, radiation therapy is limited by the sensitivity of the lesion to radiation and by the dose-dependent side effects such as cognitive deficits, leukomalacia, necrosis, and pituitary and hypothalamic dysfunction.

Oncology research is always striving to develop a tumor-specific treatment for cancer as an alternative to chemotherapy. Such an approach would target cancer cells specifically without damaging normal adjacent tissue.[16] By selectively targeting overexpressed antigens on tumor cells and having a selectivity as high as 200,000-fold compared to the native toxin, immunotoxins represent a highly selective alternative therapy.[17,18] Immunotoxins were first shown to be potent tumor cell killers in the early 1970s; however, it was not until 1987 that there were any published results of the effects of immuntoxins on primary CNS tumors.[19] Since then, 12 clinical studies assessing the safety and efficacy of immunotoxin use in primary CNS tumor treatment have been published (Table 5.1). Although these trials have shown some efficacy in the use of immunotoxins to treat CNS tumors, the trials have not shown the dramatic response of treating tumors with immunotoxins as demonstrated in *in vitro* and *in vivo* models. Two obstacles that need to be overcome for immuntoxins to become standard treatment for brain tumors are their nonuniform distribution within tumors and nonspecific treatment-related neural toxicity. Future direction in immunotoxin clinical trials will undoubtedly involve the use of immunotoxin combinations with affinities for various antigens expressed either on the tumor cells, the surrounding neural matrix, or the neovasculature. In this manner, the dose-dependent toxicity of the individual immunotoxins can be reduced without

Table 5.1 CNS immunotoxin clinical trials

Author	Immuntoxin	Adminstrative route	Tumor type	Primary or recurrent	# with resection prior to treatment	Outcome	Adverse effects
Riva et al.	Anti-Tenascin-I^{131} Mab	Intratumoral	58 GBM, 4 AA	31 primary 31 recurrent	62	23 months median survival 12 months median relapse	Headache, HAMA
Laske et al.	Tf-CRM107	Intratumoral	9 GBM, 5 AA, 1 AO	15 recurrent	11	9/15 patients with ≥ 60% tumor reduction	Seizure, cerebral edema, peritumoral cerebral injury
Wersäll et al.	Anti-EGFR Mab	Intratumoral	8 GBM	4 primary 4 recurrent	5	6 Patients with evidence of tumor necrosis	Nephropathy, headache, ataxia, hemiparesis, septicemia, HAMA, nausea/vomiting
Brady et al.	Anti-EGFR-I^{125} Mab	intravenous/ intraarterial	15 GBM, 10 AA	25 primary	12	15.6 months median survival	Nonreported
Stragliotti et al.	Anti-EGFR Mab	Intravenous	16 GBM	16 recurrent	12	No major response	Neutropenia, hepatitis, skin rash
Rand et al.	IL-4(38–37)-PE38KDEL	Intratumoral	9 GBM	9 recurrent	9	6/9 patients with evidence of tumor necrosis	Headache, seizure, dysphasia, hydrocephalus, weakness
Ascher et al.	NBI-3001	Intratumoral	31 malignant gliomas	31 recurrent	31	8.2 months median survival	Cerebral edema, seizure, aphasia, hemiparesis, coma, somnolence

Table 5.1 Continued

Author	Immuntoxin	Adminstrative route	Tumor type	Primary or recurrent	# with resection prior to treatment	Outcome	Adverse effects
Lang et al.	IL13-PE38QQR	Intratumoral	9 GBM, 1 AA, 1 AO, 1 MO	12 recurrent	12	1+ to 42+ weeks	cerebral edema, seizure, cranial nerve palsy, sensory neuropathy, ataxia
Laske et al.	454A12-rRA	Intraventricular	8 leptomeningeal neoplasia	8 primary	NA	4/8 patients with ≥ 50% reduction in CSF tumor cell count	Headache, mental status change, vomiting
Cokgor et al.	81C6 Anti-Tenascin-I[131] MAb	Intratumoral	32 GBM, 3AA, 5 AO, 2 MO	42 primary	42	79 weeks median survival	Hemiparesis, dysphasia, ataxia, memory loss, neutropenia, thrombocytopenia
Snelling et al.	Anti-EGFR-I[125] MAb	Intravenous	46 GBM, 13 AA	59 primary	48	13.5 months median survival	Nonreported
Kalofonos et al.	Anti-EGFR-I[131] Mab Anti-H17E2-I[131] MAb	Intravenous/ intraarterial	7 GBM, 1 AA, 2 brainstem gliomas	10 recurrent	8	42 days to 3 year survival rate	Thrombocytopenia, neutropenia, HAMA

GBM, glioblastoma multiforme; AA, anaplastic astrocytoma; AO, anaplastic oligodendroglioma; MO, mixed oligodendroglioma; HAMA, Human anti-mouse Antibody; NA, not applicable.

compromising their efficacy. In addition, combinations of various delivery techniques such as intratumoral convection-enhanced delivery (CED), intraarterial injections following BBB disruption, and intraventricular administration can be used to attack the tumor cells at the margin. Alternatively the use of biological response modifiers such as interferon[20] to increase the density of the antigenic sites on tumor cells or the intratumoral co-injection of hyaluronidase to facilitate the distribution of immunotoxins[21] before treatment with immunotoxin can decrease the toxicity to surrounding cells. The addition of substances such as chloroquine to decrease the toxicity of transferrin-bound immunotoxins can also be employed.[22] Finally, the use of immunotoxins that are enzymatically activated after they have bound to the tumor antigen, should reduce the potential for systemic toxicity.

Numerous tumor antigen candidates have been targeted by immunotoxins. Although a large amount of important work has been published in the field (as reviewed in this book), in this chapter we focus on a comparison of two agents produced in our laboratory, DTAT and $DT_{390}IL13$.[23,24]

IL-13 receptor

Targeting the IL-13 receptor with immunotoxin has proven already to have tremendous potential for treating brain tumors (see previous chapters in this volume). IL-13 is a cytokine that is structurally similar to IL-4. It is secreted by activated type-2 T-cells and mast cells.[25–27] Like IL-4, IL-13 modulates human monocyte and B-cell function, but unlike IL-4 it does not have biological action on T-cells.[28,29] The IL-13 receptor (IL-13R) is composed of two subunits: IL-13Rα1 or IL-13Rα2 and IL-4Rα.[30] IL-13Rα1 and IL-13Rα2 have a 45% homology in structure. Both proteins have short cytoplasmic domains, four conserved cysteine residues, two consensus patterns, and WSXWS motif (signature of the hemopoietic cytokine receptor family) in the C-terminal domain.[31]

IL-13Rα2 has been found to be overexpressed in cultured human GBM cell lines and surgical GBM specimens but not in normal human brain,[32,33] whereas IL-13Rα1 is detected in keratinocytes, hair follicles, sebaceous and sweat glands, foveola cells, gastric glands, and hepatocytes.[34] IL-13Rα1 does not transduce a signal when bound to IL-13, but the heterodimerization of IL-13Rα1 and IL-4Rα transduces a signal when either IL-4 or IL-13 is bound.[30] Unlike IL-4 receptor, IL-13R is not species-specific.[35] Although IL-13Rα2 has three orders of greater affinity for IL-13 than does IL-13Rα1, it acts as a decoy in the sense that it does not transduce a signal due to its very short cytoplasmic domain.[30,36,37]

The heterodimerization of IL-13Rα1 and IL-4Rα activates Tyk2, a member of the Janus tyrosine kinase (Jak) family, STAT 3 and STAT 6, which are members of a family of molecules known as signal transducers and activators of transcription.[38–40] Phosphorylation of STAT 6 results in its translocation to the nucleus where specific genes are upregulated. Research has indicated that the upregulated genes play a role in the pathogenesis of allergic diseases such as bronchial asthma, atopic dermatitis, allergic rhinitis, and allergic conjunctivitis by type 1 or immediate hypersensitivity reactions involving IgE synthesis.[34] Furthermore, overexpression of IL-13 can induce allergic reactions independent of type 1 hypersensitivity.[41]

In terms of immunotoxin therapy, IL-13R has been shown to be overexpressed on many human malignancies including GBM, renal cell carcinoma, ovarian carcinoma, colon adenocarcinoma, epidermoid carcinoma, AIDS-associated Kaposi sarcoma, prostate cancer, and pancreatic carcinoma.[33,42–46]

uPA receptor

The hallmark of a neoplasm transitioning from a benign tumor to malignant cancer is its ability to invade local tissue and to also metastasize. Various proteases (such as metalloproteases, type 4 collagenases, serine proteases, and plasmin-related enzymes) have been shown to play crucial roles in this initial step.[47–49] Within the past few years attention has turned to the urokinase-type plasminogen activator receptor (uPAR) as a target for inhibiting tumor cell metastass.[24,50,51] In accordance with this, uPAR has been shown to be overexpressed in a number of tumor cell lines including: GBM, breast cancer, melanoma, colon cancer, and prostate cancer.[24,52–56] In addition to its function in the plasminogen activating cascade, deficiency of uPAR has been shown to decrease T-cell and macrophage recruitment and impair transendothelial migration of neutrophils.[57–59] uPAR is a 270 residue, 55 kDa, glycosyl phosphatidylinositol (GPI)-linked protein, which is bound to the cell membrane through a GPI-linked ligand.[60] uPAR resides in low density lipoprotein domains.[61] Since GPI-linked proteins are unable to transduce a transmembrane signal, it is thought that uPAR signaling occurs through protein–protein complexes.[61] Furthermore, not all proteins that coexist in rafts form a protein–protein complex however, studies have been done which show that uPAR does function through such a pathway. Resonance energy transfer has been observed between uPAR and several integrins indicating that such complexes do exist.[62] Other eloquent experiments have shown a disruption of uPAR–integrin interaction by using a peptide with a homologous sequence to a suspected uPAR-binding site.[63,64] Recent studies indicate that the α2-macroglobulin receptor, LDL-related receptor protein, very low density lipoprotein receptor, and the epithelial glycoprotein-330 are involved in the internalization of uPAR.[65,66–69] uPAR consists of three heavily glycosylated domains, which are connected by interdomain peptides and disulfide bonds.[70–72] Domain 1 is the amino-terminal portion that functions as the binding site for uPA and also provides the catalytic site for the plasminogen cascade. Furthermore, domain 1 participates in a limited autocleavage which exposes a sequence between domain 1 and 2 that in turn results in the chemotaxis property of the uPAR. The carboxyl terminal (domain 3) allows for anchoring of the GPI, which in turn attaches the receptor to the plasma membrane. Several models have been proposed for the various functions of the uPAR. In tumors the binding of uPA to uPAR results in a complex formation with α5α1 which then binds to fibronectin. Through a caveolin-independent mechanism, a signal cascade is initiated, which activates the mitogen-activated protein kinase (Mek)-extracellular regulated kinase (ERK) pathway,[61,73] and simultaneously inhibits the growth suppressive pathway p38[MAPK] (stress-activated protein kinase 2). This inhibition leads to proliferation of tumor cells. In kidney epithelial cells, monocytes, and vascular smooth muscle cells, a two-domain form of uPAR (devoid of domain 1), which does not bind uPA, forms a complex with membrane integrins and in a caveolin-dependent manner activates Src Family kinases, which then activates focal adhesion kinase (FAK) and Fyn, a member of the nonreceptor tyrosine kinase family. This leads to ERK activation, which combined with FAK leads to matrix-dependent migration.[61,70,74,75] Tumor cell migration results from uPA-dependent binding of uPAR, which activates FAK and Src phosphorylation leading to activation of the Shc-Ras-Mek-ERK pathway.[61] The myosin light chain kinase (MLCK) is activated and phosphorylates the myosin regulatory light chain and initiates cytoskeleton contraction and cell migration.[76]

Plasminogen activator inhibitor (PAI)-1 inhibits uPA–uPAR activity by inducing internalization and degradation of uPA while recycling the free uPAR to the cell surface.[77] Furthermore, PAI-1 inhibits the binding of vitronectin to uPAR and inhibits cell adhesion and migration.[78,79]

Immunotoxin design

The accessibility of these two glioblastoma-associated receptors caused us to consider both as targets for IT therapy. Thus, we used established procedures to design hybrid immunotoxins with IL-13 or the amino terminal fragment of urokinase as the ligand moieties and truncated DT as the toxin moiety. Recombinant DTAT (diphtheria toxin amino terminal fragment of urokinase-type plasminogen activator) was synthesized by a technique described previously.[80] The hybrid gene was constructed by the method of gene splicing. An Ncol/Xhol gene fragment was cloned by polymerase chain reaction (with a splice overlap extension encoding the 390-amino acid portion of DT$_{390}$, an EASGGPE linker, and the downstream 135-amino acid ATF from uPA) and ligated into the pET21d expression vector forming plasmid pDTAT.pET21d (Novagen, Madison, WI).[81] The DT$_{390}$ consists of 193-amino acid N-terminal A-chain, the 342-amino acid B-chain with a hydrophobic translocation enhancing region, a disulfide bond between chains A and B, with the deletion of 145-amino acid native binding region. Restriction endonuclease digestion and DNA sequencing analysis were used to verify that the hybrid gene had been cloned in frame. Once this was verified, the plasmid was transformed into the *Escherichia coli* strain BL21(DE3) (Novagen). Expression was induced and the protein was refolded and purified from inclusion bodies. The pellets were washed and inclusion bodies were collected by centrifugation. Solubilization of the inclusion body pellet was achieved by sonicating in denaturant buffer. Renaturation was initiated by a rapid dilution of the denatured protein into chilled refolding buffer at 10°C for 48 h. Ultrafiltration was performed and samples were loaded on a Q-Sepharose (Sigma-Aldrich, St. Louis, MO) ion exchange column and eluted with 1 M NaCl in 20 mM Tris (pH 7.8). The protein was diluted and subsequently applied to a Resource Q column (Pharmacia Biotech, Uppsala, Sweden) and eluted with a linear salt gradient and dialyzed. The main peak from the Resource Q column was further purified by a size-exclusion chromatography. The purity of DTAT was determined to be greater than 95% based on a standard sodium dodecyl sulfate–polyacrylamide gel electrophoresis (SDS–PAGE) analysis. $DT_{390}IL13$ was synthesized in a similar fashion with a 95% purity.

DTIL13 efficacy

Investigators have developed and studied IL-13PE[32,35,45] and we reasoned that an IL-13/DT hybrid protein also would be selective and potent. Human immunization to DT would not be a problem because of use of the protein in the brain, an immunologically priviledged site. To determine the potency of $DT_{390}IL13$, various cell lines were cultured in the presence of increasing concentrations of $DT_{390}IL13$. Cells were pulsed with [methyl-^{3}H] thymidine and the radioactivity was counted. Human GBM cell lines U373 MG, U87 MG, T98 G, and control cell lines C1498 (murine leukemia) and HUT-102 (human T-cell lymphoma), which do not express IL-13R, were examined. Previous work by Debinski et al.[32,82] had determined that U373 MG had ≈ 16, 400 IL-13R/cell whereas T98 G had only 549 receptors/cell, and U87 MG was intermediate in receptor density. The IC_{50} correlated directly with the receptor density with 12 pM for $DT_{390}IL13$ and 1 nM for U87 MG cells. Higher concentrations of $DT_{390}IL13$ had minimal or no effect on the control cells and T98 G. A kinetic study of $DT_{390}IL13$ against U373 MG cells revealed greater than 50% cell death with concentrations less than 0.1 nM. $DT_{390}IL13$ also was tested against primary explant cells originating from three patients with GBM. All three samples were sensitive to $DT_{390}IL13$ in nM concentrations. To assess the specificity of $DT_{390}IL13$, U373 MG cells were incubated with $DT_{390}IL13$, $DT_{390}IL2$ and $DT_{390}mIL4$ (mouse IL-4 construct). Whereas $DT_{390}IL13$ was highly toxic, $DT_{390}IL2$ and $DT_{390}mIL4$ showed no cytotoxicity. Furthermore, while polyclonal anti-human IL-13 antibody at 1.0 µg/ml effectively neutralized the cytotoxicity of $DT_{390}IL13$, co-incubation with antihuman IL-4R or antimurine IL-4 antibody had no effect. To test the efficacy of $DT_{390}IL13$ *in vivo*, a nude mouse tumor flank model was used. U373 MG cells were inoculated subcutaneously into the right flank of nude mice. On day 12 the tumors (ranging from 0.1 to 0.25 cm^3) were treated every other day with intratumoral injections of either $DT_{390}IL13$, $DT_{390}hIL2$, or PBS for a total of five doses. All tumors treated with $DT_{390}IL13$ had complete regression ($p < 0.005$) whereas tumors injected with nonspecific immunotoxin or saline continued to grow.[20]

Although tumor flank models can provide further evidence of immunotoxin efficacy, the model has several drawbacks compared to an intracranial tumor model. First, flank models are known to autoinfarct once a critical mass threshold has been reached. Second, the intratumoral pressure within the flank model does not reach the pressure that occurs in an intracranial tumor due to the fact that the skull has a fixed volume. Third, flank xenograft models can elicit an immune response to the tumor cells whereas the intracranial model is privileged because of the blood-brain barrier.[83] Fourth, the nonspecific toxicity that can occur with immunotoxin treatment such as vascular leak syndrome can be deleterious in an intracranial model thus limiting the dose of immunotoxin that can be safely injected.[84,85] Finally, human GBM cells transplanted in nude mice brain show an infiltrative pattern of growth similar to the human pathology; whereas, similar cell transplants into mice flank do not infiltrate adjacent stroma.[86] We have developed an intracranial model using U373 MG cells to test the efficacy of $DT_{390}IL13$. Athymic nude mice were anesthetized and placed in a stereotactic frame. The scalp was cleansed and a midline incision was made extending from the bregma to the lambda. A power drill was used to perforate the skull on the right approximately 2 mm lateral to the midpoint between the bregma and the lambda. A stereotactically guided Hamilton needle was passed approximately 3.5 mm into the parenchyma to deliver the tumor cells and the scalp was sutured closed. Magnetic resonance imaging (MRI) was used to assess tumor size and posttreatment efficacy in athymic nude mice. Figure 5.1 shows a histology brain section in which U373 cells were injected intracranially and the tumor was visualized using H and E staining on day 30 post-injection. In the left panel, the tumor margins and needle tract can clearly be observed. In order to create an

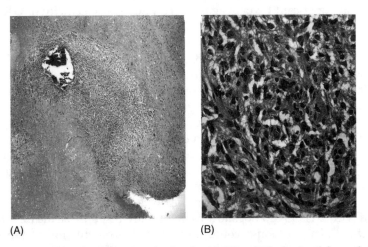

(A)   (B)

Figure 5.1 Histology brain section (stained with H and E) obtained from the putamen of nude mice given a single injection of U373 MG glioblastoma cells. Panel A shows tumor with the needle track visible in the upper left margin 200×. Panel B shows active mitosis and hypercellularity within the tumor 400×

in vivo model that approximated the clinical condition, the mice were treated at a point in time when only a week of life would be expected based on the elucidated natural history of the intracranial tumor model. After $DT_{390}IL13$ treatment, an MRI with contrast was obtained, which showed significant tumor shrinkage. The survival of the treated group was statistically significant ($p < 0.05$) compared with the normal saline-treated control group. Although the efficacy of $DT_{390}IL13$ against U373 MG tumors was evident, the fact that tumor cells in the periphery were still present on posttreatment scans indicates that a CED of $DT_{390}IL13$ will be required in future clinical trials.

DTAT efficacy

The *in vitro* cytotoxicity of DTAT was measured by inhibition of DNA synthesis in various cell lines including U87 MG, U118 MG, Neuro-2a (murine neuroblastoma cell line), Daudi (human Burkitt's lymphoma cell line), and SKBR3 (human mammary gland adenocarcinoma cell line).[24] Whereas U87 MG and U118 MG were highly sensitive to DTAT, the remaining cells that do not express uPAR were not, except for Neuro-2a. Neuro-2a showed some sensitivity at higher concentrations of DTAT probably due to the uptake of uPa by the low density lipoprotein complexes on neurons and glial cell membranes.[65–69] Since uPAR is known to be upregulated in neovascularization, wound healing, and placental development, DTAT efficacy was tested against human umbilical vein endothelial cells (HUVEC). DTAT was shown to inhibit the proliferation of HUVEC in a dose-dependent manner with an IC_{50} of 2 nM. To test the specificity of DTAT against HUVEC, a 72 h proliferation assay consisting of several interleukin-based immunotoxins (both mouse- and human-based gene constructions) was completed. Only DTAT had significant toxicity against HUVEC. The specificity of DTAT was also tested by the ability of anti-urokinase antibody to block the cytotoxicity of DTAT against HUVEC. A comparison of the cytotoxicity of DTAT and $DT_{390}IL13$ was performed using U118 MG cells in a proliferation assay. U118 MG is known to coexpress IL13R and uPAR. Whereas DTAT resulted in greater than 90% cytoreduction at 48 h, only a 75% reduction was noted with $DT_{390}IL13$ ($p < 0.001$). *In vivo* efficacy of DTAT was tested using an athymic nude mice flank tumor model. Four groups of ten mice were transplanted with U118 MG cells in the flank and on day 28 when the average tumor size had reached 0.2 cm^3, the animals were treated with intratumoral injections of DTAT, DThIL2, PBS, or $DT_{390}IL13$. By day 65, 9 of 10 mice treated with DTAT were tumor free whereas mice not treated with DTAT continued to have tumor growth. One interesting finding was that U118 MG flank tumors treated with DTIL13 did not show as favorable a response as the mice treated with DTAT. This response could be due to downregulation of the IL-13R *in vivo*. Wen et al.,[87] have shown that receptor density on tumor cells can vary depending on the micro-environment near the tumor cells. Another explanation for this result could be the poor diffusion of $DT_{390}IL13$ through the tumor bed.

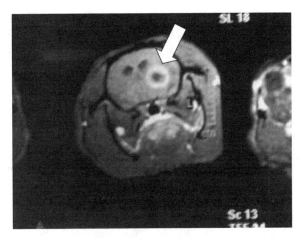

Figure 5.2 TI-weighted post-contrast MRI scan of nude mouse brain showing U87 tumor growth with central necrosis (arrow)

Recently, we completed an intracranial tumor model study assessing the efficacy of DTAT on U87 MG cells. Figure 5.2 shows an MRI scan with contrast that demonstrated the tumor growth of the U87 MG glioblastoma in the putamen of a nude mouse. DTAT administration resulted in a statistically significant ($p < 0.0001$) increase in the survival of the mice compared to controls. In order to screen for intracranial bleeding or vascular leak syndrome, serial post-injection MRI scans were obtained in rats after undergoing intraparenchymal DTAT or normal saline injections. Gradient echo scans showed no intracranial bleeding at 1 h or again at 72 h post-injection. T2-weighted and FLAIR (fluid-attenuated inversion recovery) images also showed no cerebral edema.

Toxicities

To assess the nonspecific and specific systemic toxicity of $DT_{390}IL13$ and DTAT, C57BL/6 mice were injected subcutaneously with repeat doses of either $DT_{390}IL13$ or DTAT. The maximal tolerated dose (MTD) was calculated based on immunotoxin-related death. For $DT_{390}IL13$, the therapeutic window was determined to be between 1 and 30 μg/dose. For DTAT, the MTD was 20 μg/dose. Frozen tissue specimens were stained with hematoxylin and eosin and examined microscopically. Few neutrophil and mononuclear cell infiltrates were detected in kidney glomeruli in both DTAT- and $DT_{390}IL13$-treated mice. However, serum analysis indicated normal levels of creatinine indicating no kidney dysfunction was present. $DT_{390}IL13$ did result in a dose-dependent liver damage. To evaluate whether this toxicity was specific or a bystander effect, IL4R knock-out mice were tested since IL-13R requires IL-4R to be functional [88]. Liver specimens from IL-4R knock-out mice treated with $DT_{390}IL13$ showed no fatty liver damage, and their liver function enzymes were not significantly increased. Although DTAT also showed mild elevation in liver enzymes, this result did not correlate with any histological changes in the liver structure.

Does DTAT have an advantage over DTIL13?

As far as their ability to kill target cells *in vitro* and their ability to inhibit the growth of established tumors, both DTAT and $DT_{390}IL13$ are highly effective and do not appear to show major differences. However, one important distinction is the ability of DTAT to target the tumor neovasculature rather than only overexpressed receptors on the surface of tumor cells. At this stage, it is not possible to know whether vascular targeting is a welcome side benefit or an undesirable risk. We can anticipate that DTAT will kill endothelial cells in the tumor vasculature and spare normal endothelial cells because immunohistochemical studies revealed that uPAR is mostly expressed in tissues undergoing extensive remodeling, e.g., in trophoblast cells in the placenta, in keratinocytes at the edge of incisional wounds, and in cells of primary tumor and metastasis.[89] Also, uPAR expression may be differentially higher in the proliferating endothelium. Conversely, targeting the tumor vascular in the brain may enhance the risk of intracranial bleeding and subsequent clot formation. Another advantage may be that DTAT and $DT_{390}IL13$ successfully target two unrelated receptors. Studies show[23] that IL-13R expression varies on glioblastoma lines and we have observed that some lines are inhibited more by DTAT and others are inhibited more by $DT_{390}IL13$. We expect that the same will be largely true for the treatment of clinical brain tumors. Tumors that express high levels of IL13R will respond well to $DT_{390}IL13$ and those that do not express high levels will not respond well. In these cases, it will be helpful to have an agent such as DTAT that targets a totally unrelated marker, one that appears to be expressed in high copy number on glioblastoma cells. In fact, it is possible that we may derive an additive effect of the two drugs when used in combination *in vivo*.

Future direction

Although very informative, we have reached the limitations of what we can learn in the flank tumor experiments in mice. Having demonstrated the efficacy of DTAT and $DT_{390}IL13$ *in vitro* and *in vivo*, future experiments will focus on establishing the MTD and therapeutic index in the intracranial mouse model that we have reported here. Undoubtedly, this different route of inoculation will have an impact on the efficacy and toxicity of these agents, particularly DTAT because of its vascular affects. It will also be important to perform pharmacokinetic analysis and dose escalation studies in primates to understand the dosing and dose scheduling required for the optimum effect of this agent.

References

1. Hall WA, Fodstad Ø: Immunotoxins and central nervous system neoplasias. *J Neurosurg* 76:1–12, 1992.
2. Kirby S, Macdonald D, Fisher B, et al.: Preradiation chemotherapy for malignant glioma in adults. *Can J Neurol Sci* 23:123–127, 1996.
3. Kristensen CA, Kristjansen PE, Hansen HH: Systemic chemotherapy of brain metastases from small-cell lung cancer: a review. *J Clin Oncol* 10:1498–1502, 1992.
4. McAllister LD, Doolittle ND, Gustadisegni PE, et al.: Cognitive outcomes and long-term follow up after enhanced chemotherapy delivery for primary central nervous system lymphomas. *Neurosurgery* 46:51–61, 2000.
5. Walker MD, Green SB, Byar DP, et al.: Randomized comparisons of radiotherapy and nitrosureas for the treatment of malignant glioma after surgery. *N Engl J Med* 303:1323–1329, 1980.
6. Salcman M: The morbidity and mortality of brain tumors. A perspective on recent advances in therapy. *Neurol Clin* 3:229–257, 1985.
7. Nazarro JM, Neuwalt EA: The role of surgery in the management of supratentorial intermediate and high-grade astrocytomas in adults. *J Neurosurg* 73:331–344, 1990.
8. Greene GM, Hitchon PW, Schelper RL, Yuh W, Dyste GN: Diagnostic yield in CT-guided stereotactic biopsy of gliomas. *J Neurosurg* 71:494–497, 1989.
9. Kelly PJ, Daumas-Duport C, Kispert DB, Kall BA, et al.: Imaging-based stereotactic serial biopsies in untreated intracranial glial neoplasms. *J Neurosurg* 66:865–874, 1987.
10. Burger PC, Heinz ER, Shibata T, Kleihues P: Topographic anatomy and CT correlations in the untreated glioblastoma multiforme. *J Neurosurg* 68:698–704, 1988.
11. Bergstrom M, Collins P, Ehrin E: Discrepancies in brain tumor extent as shown by computed tomography and positron emission tomography using [^{68}Ga] EDTA, [^{11}C]glucose, and [^{11}C] methionine. *J Comput Assist Tomogr* 7:1062–1066, 1983.
12. Earnest F, Kelly PJ, Scheithauer BW, Kall BA, et al.: Cerebral astrocytomas: histopathologic correlation of MR and CT contrast enhancement with stereotactic biopsy. *Radiology* 166:823–827, 1988.
13. Kroll RA, Neuwelt EA: Outwitting the blood-brain barrier for therapeutic purposes: osmotic opening and other means. *Neurosurgery* 42(5):1083–1099, 1998.
14. Skipper H: Historic milestones in cancer biology: a few that are important to cancer treatment (revised). *Semin Oncol* 6:506–514, 1979.
15. Morrow C, Cowan K: Drug resistance and cancer. *Adv Exp Med Biol* 330:287–305, 1993.
16. Ehrlich P: The relationship between chemical constitution, distribution, and pharmacological action. In: Himmelweit F, Marquardt M, Dale H (eds.) *The Collected Papers of Paul Ehrlich, Vol. 1.* Pergamon Press, Elmsford, NY, 1956, pp. 596–618.
17. Siegall CB: Targeted toxins as anticancer agents. *Cancer* 74(3):1006–1012, 1994.
18. Hall WA: Targeted toxin therapy. In: Korblith PL, Walker MD (eds.) *Advances in Neuro-Oncology II.* Futura Armonk, NY, 1997, pp. 505–516.

19. Zovickian J, Johnson VG, Youle RJ: Potent and specific killing of human malignant brain tumor cells by an anti-transferrin receptor antibody-ricin immunotoxin. *J Neurosurg* 66:850–861, 1987.

20. Jain RK: Delivery of novel therapeutic agents in tumors: physiological barriers and strategies. *JNCI* 81:570–576, 1989.

21. Swabb EA, Wei J, Gullino PM: Diffusion and convection in normal and neoplastic tissues. *Cancer Res* 34:2814–2822, 1974.

22. Hagihara N, Walbridge S, Olson AW, et al.: Vascular protection by chloroquine during brain tumor therapy with Tf-CRM107. *Cancer Res* 60:230–234, 2000.

23. Li C, Hall WA, Jin N, et al.: Targeting glioblastoma multiforme with an IL-13/diphtheria toxin fusion protein *in vitro* and *in vivo* in nude mice. *Protein Eng* 15(5):419–427, 2002.

24. Vallera DA, Li C, Jin N, et al.: Targeting urokinase-type plasminogen activator receptor on human glioblastoma tumors with diphtheria toxin fusion protein DTAT. *J Natl Cancer Inst* 94(8):597–606, 2002.

25. Brown KD, Zurawski SM, Mosmann TR, et al.: A family of small inducible proteins secreted by leukocytes are members of a new superfamily that includes leukocyte and fibroblast-derived inflammatory agents, growth factors, and indicators of various activating processes. *J Immunol* 15:679–687, 1989.

26. McKenzie AN, Culpepper JA, de Waal Malefyt R, et al.: Interleukin 13, a T-cell-derived cytokine that regulates human monocyte and B-cell function. *Proc Natl Acad Sci USA* 90:3735–3739, 1993.

27. Minty A, Chalon P, Derocq JM, et al.: Interleukin 13 is a new human lymphokine regulating inflammatory and immune responses. *Nature* 362:248–250, 1993.

28. Hilton DJ, Zhang JG, Metcalf D, et al.: Cloning and characterization of a binding subunit of the interleukin 13 receptor that is also a component of the interleukin 4 receptor. *Proc Natl Acad Sci USA* 93:497–501, 1996.

29. Zurawski G, de Vries JE: Interleukin 13, an interleukin 4-like cytokine that acts on monocytes and B cells, but not T cells. *Immunol Today* 15:19–26, 1994.

30. Miloux B, Lament P, Bonnin O, et al.: Cloning of the human IL-13Rα1 chain and reconstitution with the IL-4Rα of a functional IL-4/IL-13 receptor complex. *FEBS Lett* 401:163–166, 1997.

31. Kishimoto T, Taga T, Akira S: Cytokine signal transduction. *Cell* 76:253–262, 1994.

32. Debinski W, Obiri NI, Powers SK, et al.: Human glioma cell lines over express receptors for interleukin 13 and are extremely sensitive to a novel chimeric protein composed of Interleukin 13 and Pseudomonas exotoxin. *Clin Cancer Res* 1:1253–1258, 1995.

33. Debinski W, Gibo DM, Slagle B, et al.: Receptor for interleukin 13 is abundantly and specifically over-expressed in patients with glioblastoma multiforme. *Int J Oncol* 15(3):481–486, 1999.

34. Akaiwa M, Yu B, Umeshita-Suyama R, et al.: Localization of human interleukin 13 receptor in non-haematopoietic cells. *Cytokine* 13(2):75–84, 2001.

35. Puri RK, Hoon DS, Leland P, et al.: Preclinical development of a recombinant toxin containing circularly permuted interleukin 4 and a truncated Pseudomonas exotoxin for therapy of malignant astrocytomas. *Cancer Res* 56:5631–5637, 1996.

36. Caput D, Laurent P, Kaghad M, et al: Cloning and characterization of a specific interleukin (IL)-13 binding protein structurally related to the IL-5 receptor alpha chain. *J Biol Chem* 271:16921–16926, 1996.

37. Donaldson DD, Whitters MJ, Fitz LJ, et al.: The murine IL-13 receptor alpha 2: molecular cloning, characterization, and comparison with murine IL-13 receptor alpha 1. *J Immunol* 161:2317–2324. 1998.

38. Izuhara K, Umeshita-Suyama R, Akaiwa M, et al.: Recent advances in understanding how interleukin-13 signals are involved in the pathogenesis of bronchial asthma. *Arch Immunol Ther Exp* 48:502–512, 2000.

39. Umeshita-Suyama R, Sugimoto R, Akaiwa M, et al.: Characterization of IL-4 and IL-13 signals dependent on the human IL-13 receptor alpha chain 1: redundancy of requirement of tyrosine residue for STAT 3 activation. *Int Immunol* 12:1499–1509, 2000.

40. Darnell JE: STATs and gene regulation. *Science* 277:1630–1635, 1997.

41. Corry DB: IL-13 in allergy. *Curr Opin Immunol* 11:610–614, 1999.

42. Obiri NI, Debinski W, Leonard WJ, et al.: Receptor for interleukin 13. Interaction with interleukin 4 by a mechanism that does not involve the common gamma chain shared by receptors for interleukin 2,4,7,9,and 15. *J Biol Chem*, 270:8797–8804, 1995.

43. Obiri NI, Leland P, Murata T, et al.: The Il-13 receptor structure differs on various cell types and may share more than one component with IL-4 receptor. *J Immunol* 158:756–764, 1997.

44. Murata T, Obiri NI, Puri RK: Human ovarian-carcinoma cell lines express IL-4 and IL-13 receptors: comparison between IL-4 and IL-13 induced signal transduction. *Int J Cancer* 70:230–240, 1997.

45. Debinski W, Obiri NI, Puri RK: A novel chimeric protein composed of interleukin 13 and Pseudomonas exotoxin is highly cytotoxic to human carcinoma cells expressing receptors for interleukin 13 and interleukin 4. *J Biol Chem* 270:16775–16780, 1995.

46. Kornmann M, Kleef J, Debinski W, et al.: Pancreatic cancer cells express interleukin-13 and interleukin-4 receptors and their growth is inhibited by Pseudomonas exotoxin coupled to interleukin-13 and -4. *Anticancer Res* 19:125–131, 1999.

47. Woolley DE: Collagenolytic mechanisms in tumor cell invasion. *Cancer Metast Rev* 3:361–372, 1984.

48. Reich R, Thompson EW, Iwamoto YY, et al.: Effects of inhibitors of plasminogen activator, serine proteinases, and collagenase IV on the invasion of basement membranes by metastatic cells. *Cancer Res* 48:3307–3312, 1988.

49. Blasi F: Urokinase and urokinase receptor: a paracrine/autocrine system regulating cell migration and invasiveness. *BioEssay* 15:105–111, 1993.

50. Fazioli F, Blasi F: Urokinase-type plasminogen activator and its receptor: new targets for anti-metastatic therapy? *Trends Pharm Sci* 15:25–29, 1994.

51. Chapman HA: Plasminogen activators, integrins, and the coordinated regulation of cell adhesion and migration. *Curr Opin Cell Biol* 9:714–724, 1997.

52. Grøndahl-Hansen J, Peters HA, van Putten WLJ, et al.: Prognostic significance of the receptor for urokinase plasminogen activator in breast cancer. *Clin Cancer Res* 1:1079–1087, 1995.

53. Christensen L, Wiborg-Simonsen AC, Heegaard CW, et al.: Immunohistochemical localization of urokinase-type plasminogen activator type-1, plasminogen-activator inhibitor, urokinase receptor and alpha(2)-macroglobulin receptor in human breast carcinomas. *Int J Cancer* 66:441–452, 1996.

54. de Vries T, Mooy C, Van Balken M, et al.: Components of the plasminogen activation system in uveal melanoma a clinico-pathological study. *J Pathol* 175:59–67, 1995.

55. Verspaget HW, Sier CF., Ganesh S, et al.: Prognostic value of plasminogen activators and their inhibitors in colorectal cancer. *Eur J Cancer* 31A:1105–1109, 1995.
56. Crowley CW, Cohen RL, Lucas BK, et al.: Prevention of metastasis by inhibition of the urokinase receptor. *Proc Natl Acad Sci USA* 90:5021–5025, 1993.
57. Gyetko MR, Chen GH, McDonald RA, et al.: Urokinase is required for the pulmonary inflammatory response to Cryptococcus neoformans. A murine transgenic model. *J Clin Invest* 97:1818–1826, 1996.
58. Pedersen TL, Yong K, Pedersen JO, et al.: Impaired migration *in vitro* of neutrophils from patients with paroxysmal nocturnal haemoglobinuria. *Br J Haematol* 95:45–51, 1996.
59. Bianchi E, Ferrero E, Fazioli F, et al.: Integrin-dependent induction of functional urokinase receptors in primary T-lymphocytes. *J Clin Invest* 98:1133–1141, 1996.
60. Blasi F, Vassalli JD, Dano K: Urokinase-type plasminogen activator: proenzyme, receptor, and inhibitors. *J Cell Biol* 104:801–804, 1987.
61. Ossowski L, Aguirre-Ghiso JA: Urokinase receptor and integrin partnership: coordination of signaling for cell adhesion, migration, and growth. *Curr Opin Cell Biol* 12:613–620, 2000.
62. Xue W, Mizukami I, Todd RF III, et al.: Urokinase-type plasminogen activator receptors associate with beta 1 and beta 3 integrins of fibrosarcoma cells: dependence on extracellular matrix components.*Cancer Res* 57:1682–1689, 1997.
63. Wei Y, Lukashev M, Simon DI, et al.: Regulation of integrin function by the urokinase receptor. *Science* 273:1551–1555, 1996.
64. Simon DI, Wei Y, Zhang L, et al.: Identification of a urokinase receptor-intergrin interaction site. Promiscuous regulator of integrin function. *J Biol Chem* 275:10228–10234, 2000.
65. Andreasen PA, Sottrup-Jensen L, Kjoller L, et al.: Receptor-mediated endocytosis of plasminogen activators and activator/inhibitor complexes. *FEBS Lett* 338:239–245, 1994.
66. Heegaard CW, Simonsen AC, Oka K, et al.: Very low density lipoprotein receptor binds and mediates endocytosis of urokinase-type plasminogen activator-type-1 plasminogen activator inhibitor complex. *J Biol Chem* 270:20855–20861, 1995.
67. Argraves KM, Battey FD, MacCalman CD, et al.: The very low density lipoprotein receptor mediates the cellular catabolism of lipoprotein lipase and urokinase-plasminogen activator inhibitor type 1 complexes. *J Biol Chem* 270: 26550–26557, 1995.
68. Moestrup SK, Nielsen S, Andreasen P, et al.: Epithelial glycoprotein-330, mediates endocytosis of plasminogen activator-plasminogen activator inhibitor type-1 complexes. *J Biol Chem* 268:16564–16570, 1993.
69. Willnow TE, Goldstein JL, Orth K, et al.: Low density lipoprotein receptor-related protein and gp330 bind similar ligands, including plasminogen activator-inhibitor complexes and lactoferrin, an inhibitor of chylomicron remnant clearance. *J Biol Chem* 267:26172–26180, 1992.
70. Fazioli F, Resnati M, Sidenius N, et al.: A urokinase-sensitive region of the human urokinase receptor is responsible for its chemotactic activity. *EMBO J* 16:7279–7286, 1997.
71. Behrendt N, Ronne E, Dano K: Domain interplay in the urokinase receptor. Requirement for the third domain in high affinity ligand binding and

demonstration of ligand contact sites in distinct receptor domains. *J Biol Chem* 271:22885–22894, 1996.

72. Ploug M, Rahbek-Nielsen H, Nielson PF, et al.: Glycosylation profile of a recombinant urokinase-type plasminogen activator receptor expressed in Chinese hamster ovary cells. *J Biol Chem* 273:13933–13943, 1998.

73. Aguirre-Ghiso JA, Kovalski K, Ossowski L: Tumor dormancy induced by downregulation of urokinase receptor in human carcinoma involves integrin and MAPK signaling. *J Cell Biol* 147:89–104, 1999.

74. Wei Y, Yang X, Liu Q, et al.: A role for caveolin and the urokinase receptor in integrin-mediated adhesion and signaling. *J Cell Biol* 144:1285–1294, 1999.

75. Resnati M, Guttinger M, Valcamaonica S, et al.: Proteolytic cleavage of the urokinase receptor substitutes for the agonist-induced chemotactic effect. *EMBO J* 15:1572–1582, 1996.

76. Nguyen DH, Catling AD, Webb DJ, et al.: Myosin light chain kinase functions downstream of Ras/ERK to promote migration of urokinase-type plasminogen activator-stimulated cells in an integrin-selective manner. *J Cell Biol* 146:149–164, 1999.

77. Cubellis MV, Wun TC, Blasi F: Receptor-meidated internalization and degradation of urokinase is caused by its specific inhibitor of PAI-1. *EMBO J* 9:1079–1085, 1990.

78. Stefansson S, Lawrence DA: The serpin PAI-1 inhibits cell migration by blocking integrin alpha V beta 3 binding to vitronectin. *Nature* 383:441–443, 1996.

79. Køller L, Kanse SM, Kirkegaard T, et al.: Plasminogen activator inhibitor-1 repress integrin- and vitronectin-mediated cell migration independently of its function as an inhibitor of plasminogen activation. *Exp Cell Res* 232:420–429, 1997.

80. Vallera DA, Panoskaltsis-Mortari A, Jost C, et al.: Anti-graft-versus-host disease effect of DT_{390}-anti-CD3sFv, a single-chain Fv fusion immunotoxin specifically targeting the CD3 epsilon moiety of the T-cell receptor. *Blood* 88(6):2342–2353, 1996.

81. Vallera DA, Li C, Jin N, et al.: Targeting urokinase-type plasminogen activator receptor on human glioblastoma tumors with diphtheria toxin fusion protein DTAT. *J Natl Cancer Inst* 94(8):597–606, 2002.

82. Debinski W, Miner R, Leland P, et al.: Receptor for interleukin (IL) 13 does not interact with IL4 but receptor for IL4 interacts with IL13 on human glioma cells. *J Biol Chem* 271:22428–22433, 1996.

83. Parsa AT, Chakrabarti I, Hurley PT, et al.: Limitation of C6/Wistar rat intracerebral glioma model: implication for evaluating immunotherapy. *Neurosurgery* 47(4):993–999, 2000.

84. Vitetta ES: Immunotoxin and vascular leak syndrome. *Cancer* 6(3):5218–5224, 2000.

85. Baluna R, Vitetta ES: An *in vivo* model to study of immunotoxin-induced vascular leak syndrome in human tissue. *J Immunother* 22(1):41–47, 1999.

86. Antunes L, Angio-Duprez KS, Bracard SR, et al.: Analysis of tissue chimerism in nude mouse brain and abdominal xenograft models of human glioblastoma multiforme: what does it tell us about the models and about glioblastoma biology and therapy? *J Histochem Cytochem* 48(6):847–858, 2000.

87. Wen DY, Hall WA, Conrad J, et al.: *In vitro* and *in vivo* variation in trasferrin receptor expression on a human medulloblastoma cell line. *Neurosurgery* 36(6):1158–1163, 1995.

88. Murata T, Taguchi J, Puri RK: Interleukin-13 receptor alpha' but not alpha chain: a functional component of interleukin-4 receptors. *Blood* 91(10): 3884–3891, 1998.
89. Solberg H, Ploug M, Hoyer-Hansen G, Nielsen BS, Lund LR. The murine receptor for urokinase-type plasminogen activator is primarily expressed in tissues actively undergoing remodeling. *J Histochem Cytochem* 49(2):237–246, 2001.

chapter six

$DAB_{389}IL-2$ (Ontak™): Development and therapeutic applications

Paul J. Shaughnessy and Charles F. LeMaistre

Contents

Introduction

The systemic therapy of cancer has relied for many years on the nonspecific actions of cytotoxic chemicals to kill neoplastic cells. However, chemotherapy

0-4152-6365-4/05/$0.00+$1.50
© 2005 by CRC Press

is limited by collateral damage to normal tissues, which results in dose-limiting toxicities. Even when bone marrow toxicity is overcome by hematopoietic stem cell transplantation after high-dose chemotherapy, other organs, such as the heart, lungs, and liver, can develop dose-limiting toxicities. The most common cause of treatment failure after high-dose chemotherapy and autologous stem cell rescue is relapse of primary disease,[1] demonstrating that even maximal doses of cytotoxic drugs have limited effects against some malignant clones. The narrow therapeutic index of cytotoxic drugs and the ability of cancer cells to develop resistance to these drugs make it imperative to develop new mechanisms of attacking malignancies alone, or in combination with, cytotoxic agents. The rational design of novel agents in the treatment of cancer should address the systemic nature of many diseases, be selective for the target tumor cell population, and employ novel mechanisms of action. $DAB_{389}IL-2$, denileukin diftitox (Ontak™ [Ligand Pharmaceuticals Inc, San Diego, CA]) and its precursor $DAB_{486}IL-2$ are genetically engineered fusion toxin that encompass these characteristics. This chapter describes the development of $DAB_{389}IL-2$ and current treatment modalities.

Diphtheria toxin

Native diphtheria toxin (DT) is a 535-amino acid protein consisting of three domains. Fragment A is the enzymatically active domain, fragment B has a hydrophobic domain at the N-terminal portion, and the C-terminal portion of fragment B is the receptor-binding domain.[2] DT intoxicates sensitive eukaryotic cells by receptor-mediated endocytosis. Once internalized into an acidic vesicle, the enzymatically active fragment A portion is released into the cytosol. Protein synthesis is inhibited via fragment A-catalyzed adenine diphosphate (ADP) ribosylation of elongation factor 2 and ultimately results in cell death.[2,3] This is a potent and efficient process such that one molecule of diphtheria toxin can inhibit up to 2000 ribosomes/min in cell-free systems.[4]

IL-2 receptor expression

The human interleukin-2 receptor (IL-2R) can be present in low-, intermediate-, and high-affinity forms that are defined by three distinct proteins. The low-affinity receptor is defined by a 55 kD protein alone (CD25, p55, TAC, α chain) and the intermediate-affinity receptor is defined by a combination of a 75 kD (CD122, p75, β chain) and 64 kD (CD 132, p64, γ chain) protein. All three proteins together constitute the high affinity IL-2R.[5,6] Expression of the p55 alpha subunit has been reported in a variety of normal and malignant cells, but most resting T-cells, B-cells, and macrophages in the circulation do not display IL-2 receptors. Approximately 5% of freshly isolated, unstimulated human peripheral blood T lymphocytes react with anti-Tac.[7] The intermediate-affinity receptor can be constitutively expressed on natural killer cells.[5,8] The high-affinity receptor is normally restricted to activated

T- and B-lymphocytes and macrophages.[9] All or part of the IL-2R have been reported on various hematologic malignancies, including low and intermediate grade non-Hodgkin's lymphoma (NHL), Hodgkin's disease (HD), human T-cell lymphotropic virus type 1 associated adult T-cell leukemia (HTLV-1), chronic lymphocytic leukemia (CLL), and cutaneous T-cell lymphoma (CTCL).[10-13]

Development of a fusion immunotoxin

Targeting cells that expressed the IL-2R for selective cell death afforded the possibility of selective treatment of some hematologic malignancies, as well as manipulation of mediators of important immune functions. The first recombinant ligand toxin fusion protein tested *in vitro* and in animals was DAB$_{486}$IL-2. The recombinant gene was created by replacing the DT receptor-binding domain with a synthetic gene encoding IL-2 and a translational stop signal.[14] DAB$_{486}$IL-2 was expressed in *Escherichia coli* and recovered by immunoaffinity chromatography and high performance liquid chromatography. This 68 kD protein contains the first 486 amino acids of DT and the full-length sequence for human IL-2.[14,15] The protein product of this recombinant fusion gene is biologically active as a ligand that is capable of binding and initiating signal transduction events in cells expressing IL-2R.[16] The purified chimeric toxin has been shown to selectively inhibit protein synthesis in high-affinity IL-2R-bearing target cells, and to not intoxicate cells that bear either the low or intermediate form of IL-2R.[17] Approximately 10 to 40 sites are bound in an hour at optimal concentrations *in vitro*. As with native DT, internalization of only one molecule is theoretically required to cause cell death. Inhibition of protein synthesis occurs in 1 to 4 h with apoptotic cell death occurring in 1 to 4 days.[17,18]

Early clinical studies of DAB$_{486}$IL-2

In the first study of a genetically engineered ligand–fusion protein, 18 patients with hematologic malignancies were given escalating doses of drug from 0.0007 to 1.4 mg/kg/day from 1 to 7 days, each dose delivered as a bolus over 1 to 5 min.[19] The maximal tolerated dose (MTD) was 0.1 mg/kg/day for 10 doses, established by asymptomatic, reversible elevations of hepatic transaminases. Mild side effects of nausea and localized rash occurred in one patient each. Proteinuria that was reversible developed in three patients. Fever was sporadically observed in five patients, but either spontaneously resolved or responded to acetaminophen. No infections occurred that were felt to be drug related. Pharmacokinetic data showed that the serum half-life of DAB$_{486}$IL-2 was approximately 5 min and that clearance of the drug closely fit a one-compartment open mathematical model. Soluble receptors for IL-2 were present in all 18 patients and were shown to decline in 3 patients in association with a response. Before receiving DAB$_{486}$IL-2 four patients had detectable anti-DT antibodies, and two patients had detectable anti-DAB$_{486}$IL-2 antibodies. Of the 18 patients 9 were found to have antibody to both DT and DAB$_{486}$IL-2 by the end of the study period. Two patients experienced a partial response in the presence of antibodies to DT and DAB$_{486}$IL-2. One patient was given an infusion of IVIG in-between doses of DAB$_{486}$IL-2 with pharmacokinetic studies performed before and after each dose. Despite the presence of significant levels of anti-DT antibodies after the IVIG infusion, there was no effect on peak serum levels or clearance of the fusion toxin. No detectable effects on immune function were observed and no significant changes in complement levels observed in this study. Of 12 non-leukemic patients 10 had no significant changes in the absolute number of lymphocytes in the peripheral circulation. Two patients had a lymphocytosis in conjunction with DAB$_{486}$IL-2 administration with an increase in the percent and absolute number of cells that coexpress CD4 and CD25. No changes in circulating cells expressing CD16 or CD56 were observed. Antitumor effects were observed in at least one patient treated at each dose level. A patient with a follicular large cell lymphoma had a complete response that lasted for over 18 months. A patient with CLL and one with follicular small cell lymphoma each developed a partial remission lasting 5 and 12 months, respectively.

In a follow up study, the short half-life of DAB$_{486}$IL-2 was addressed by using a 90 min infusion of the drug.[20] Since optimal cell killing *in vitro* is a function of both concentration and contact time, this study was performed to determine if response rates could be improved with longer infusions. Of the 23 patients treated, 17 had hematologic diseases similar to the previous study and 3 had cutaneous T-cell lymphoma (CTCL) and 3 had Kaposi's sarcoma. A total of 51 courses of DAB$_{486}$IL-2 were delivered in a dose escalation format beginning at 0.2 mg/kg/day daily for 5 days. Toxicities were similar to the previous report and the MTD was 0.3 mg/kg/day defined by renal insufficiency associated with evidence of hemolysis, anemia, and thrombocytopenia. Seven infections were noted in five patients, however none were felt to be study related. Pharmacokinetic analysis fit a one-compartment model with a mean serum half-life of 11.5 min. Neither clearance nor area under the curve were significantly different from the earlier study. After one or more treatments 10 of 18 patients tested had low-level anti-IL-2 titers and all patients with tumor responses developed anti-IL-2 antibodies. Two patients had partial remissions, one with non-Hodgkin's lymphoma and the other with CTCL. Although it was feasible to maintain target concentrations during the infusion, a correlation with tumor response was not demonstrated.

Another study noted the response of five patients with CTCL to DAB$_{486}$IL-2.[21] One patient attained a CR sustained for over 33 months, and two others achieved PRs. A phase II study of DAB$_{486}$IL-2 was performed in 15 patients with refractory CTCL.[22] One patient had a PR with significant clearing of plaque stage disease. Two patients with Sezary syndrome had improvements in skin disease without significantly affecting circulating Sezary cells. In none of these studies was there a clinical correlation of disease response to the presence of the IL-2 receptor. These studies collectively demonstrated the premise that a rationally designed fusion toxin ligand could safely be administered to patients in bolus infusions and have some clinical activity in lymphoid malignancies. Because of the short half-life of DAB$_{486}$IL-2, as well as the complexity of its production, modifications of the molecule were made resulting in the development of a new and more potent fusion toxin ligand.

Development of DAB$_{389}$IL-2

DAB$_{389}$IL-2 differs from the earlier version, DAB$_{486}$IL-2, in the deletion of 97 amino acids from a portion of the native receptor-binding domain resulting in a lower molecular weight (57 kD VS. 68 kD, respectively). This modification also resulted in a five-fold improvement in affinity of DAB$_{389}$IL-2 for the IL-2R, approximately 10-fold increase in potency, and longer half-life compared to the earlier version.[23]

DAB$_{389}$IL-2 is a recombinant DNA-derived cytotoxic protein composed of the amino acid sequences for DT fragments A and B (Met1-Thr387)-His followed by the sequences for IL-2 (Ala1-Thr133) (Figure 6.1) and is produced in an *E.coli* expression system. The gene is carried on a genetically engineered plasmid and the protein is recovered as an inclusion body following an extensive washing procedure. DAB$_{389}$IL-2 is further purified by reverse phase chromatography followed by a multistep diafiltration process. Biodistribution and excretion studies determined that the kidneys and liver were the primary sites of accumulation outside the vasculature. DAB$_{389}$IL-2 was metabolized by proteolytic degradation with 25% of the total injected dose excreted in low molecular weight breakdown products.[24] The mechanism of action for targeted cell death is the same as described for the 486 fusion toxin and is depicted in Figure 6.2.

Clinical studies with DAB$_{389}$IL-2

Preclinical studies done with murine models of lymphoma were important predictors of observations about DAB$_{389}$IL-2 made in the clinic. For example, CP3, a cell line expressing high-affinity IL-2 receptors causes a fatal lymphoma when injected into C57BL/6 mice. These mice develop a malignancy, which causes adenopathy in 25 days and death by day 40. DAB$_{389}$IL-2 was

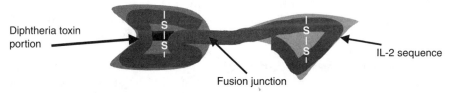

Diphtheria toxin portion

Fusion junction

IL-2 sequence

Figure 6.1 Immunotoxin fusion ligand DAB$_{389}$IL-2. DAB$_{389}$IL-2 is a recombinant DNA derived cytotoxic protein composed of IL-2 protein sequence fused to the active portion of diphtheria toxin. IL-2, interleukin 2 (Courtesy of Ligand Pharmaceuticals Inc.)

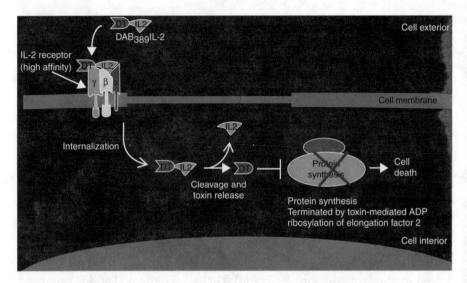

Figure 6.2   Mechanism of action of DAB$_{389}$IL-2. The fusion toxin selectively binds to cells bearing the high-affinity IL-2 receptor and is internalized resulting in cell death. DT, diphtheria toxin; IL2, interleukin 2 (Courtesy of Ligand Pharmaceuticals Inc.)

shown to delay the onset of tumors, extend median survival, and result in 80 to 100% long-term, tumor-free survival. The presence of anti-DT antibodies did increase clearance and diminish antitumor activity. This effect appeared to be overcome with increased dose.

Multi-institutional phase I/II studies evaluated dosing and pharmacokinetics of DAB$_{389}$IL-2 in humans with IL-2R expressing malignancies.[26,27] In the first study, immunostaining with an anti-CD25 antibody of greater than 25% on tumor cells was required for enrollment. Patients were enrolled in cohorts of three beginning at 3 μg/kg/day of DAB$_{389}$IL-2 in successively increasing dose levels. A treatment cycle consisted of a short infusion over 5 min daily for 5 days and was repeated every 3 to 4 weeks. No premedications were allowed except for acetaminophen and diphenhydramine. There were 230 patients screened for enrollment into this trial and 109 patients were found to have tumors meeting entry criteria for IL-2 expression: 63% expressed p55 only, 6% expressed p75 only, and 31% expressed both p55 and p75. Of the 73 patients enrolled in the study, 35 patients had CTCL, 17 had NHL, and 21 had HD. A total of 245 courses of DAB$_{389}$IL-2 were administered with the median number of courses per patient being three and 71% of patients receiving at least two courses.

Dose-limiting toxicity occurred at 31 μg/kg/day due to reversible asthenia, fever, and nausea/vomiting. Nearly all (84 of 91, 92%) of the grade 3 and 4 clinical adverse events occurred during the first or second course of treatment, with a significant decrease in subsequent courses. This observation of fewer side effects with repeated cycles has been confirmed in other studies and may correlate with the appearance of antibodies. Of 73 patients 15 (21%) had evidence of a hypersensitivity reaction manifested by wheezing, dyspnea, bronchospasm, or chest tightness. Of the 15 patients 9 experienced these symptoms only during course 1 and only one patient, with circulating malignant NHL cells, experienced symptoms severe enough to be discontinued from the study. Hypotension was reported in 40 patients and 23 patients (32%) required treatment with fluids. Six patients (8%), all with CTCL, experienced hypotension, decreased serum albumin, and edema, representing a vascular leak syndrome (VLS). The most frequent laboratory abnormalities were reversible hepatic transaminase elevations (62%) and decreased serum albumin (86%).

Pharmacokinetic parameters revealed a monophasic clearance with an overall half-life of 72 min and there was no accumulation of drug with repeated administrations. Antibody titers to DAB$_{389}$IL-2, DT, and IL-2 revealed that 38% of patients had detectable titers of antibody at baseline and 92% of patients had detectable titers after two courses (Figure 6.3). The presence of antibodies did not preclude a response and the same proportion of patients with and without a response to treatment had significant titers of antibody. Baseline data on soluble IL-2R was available in 67 patients and 37 patients (55%) had levels that were greater than 150 pmol/l. There was generally a decrease in the mean soluble IL-2R level over the first two treatment courses (Figure 6.4). Peripheral blood lymphocyte evaluations reveal there was an increase in overall T-cell expression of CD25 after DAB$_{389}$IL-2, but there was no change in the proportion of cells expressing CD3, CD4, and CD8 with CD25.

Overall, there were 16 patients who experienced tumor response. In 35 patients with CTCL there were 5 CRs and 8 PRs (RR 37%), and 3 of 17 patients

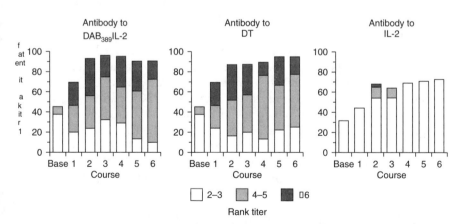

Figure 6.3   Antibody titers after administration of DAB$_{389}$IL-2. Percent of patients with rank titers for antibodies to DAB$_{389}$IL-2, DT, and IL-2 by course. Relative rank titers are shown with the following designations: rank titer 1, dilution 1:5; rank titer 2 to 3, dilution 1:25 to 1:125; rank titer 4 to 5, dilution 1:625 to 1:3125; rank titer 6, dilution 1:15625 (From LeMaistre, C.F. et al., *Blood*, 1998, 91, 399.)

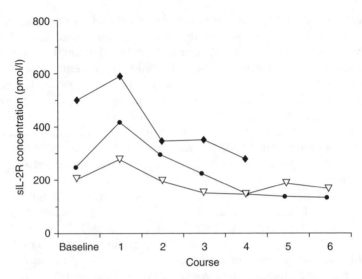

Figure 6.4 Serum IL-2R concentrations. Median serum IL-2R levels for all patients by diagnosis before and after DAB$_{389}$IL-2 administration. (▼) Hodgkin's disease; (●) cutaneous T-cell lymphoma; (▽) non-Hodgkin's lymphoma (From LeMaistre, C.F. et al., *Blood*, 1998, 91, 399.)

with NHL responded (1CR and 2 PRs). The median time to response was 2 months with a median duration of 10 months (range 2.4 to 39+). One patient with an intermediate grade NHL relapsing after autologous stem cell transplant remains in CR more than a decade after receiving DAB$_{389}$IL-2 (LeMaistre, personal communication). No patients with HD responded, however this was a very heavily pretreated group with over half of the patients having prior high-dose chemotherapy and an autologous stem cell transplant.

The most impressive responses to treatment with DAB$_{389}$IL-2 were seen in the CTCL patients as described by Saleh et al. in the same patient population.[26] A patient with extensive tumor stage CTCL had a CR, as well as another patient with diffuse erythroderma. The CTCL patients had rapid responses within the first two cycles of therapy and a median duration of response of 15 months in the patients who achieved a CR. Interestingly, two of four patients who had previously received DAB$_{486}$IL-2 responded and another patient had stabilization of disease. This clinical observation supports the increase in potency of DAB$_{389}$IL-2 compared to DAB$_{486}$IL-2.

Phase III pivotal study of DAB$_{389}$IL-2 in CTCL

CTCL is a group of diseases characterized by infiltration of the skin by mature malignant clonal lymphocytes that are usually CD4+ or CD8+ cells.[28–30] Included in this group of diseases are mycosis fungoides, Sezary syndrome, and peripheral T-cell lymphomas involving the skin.[30] Initial therapies are directed at treating the symptoms such as pruritus and recurrent skin infections with agents including mechlorethamine, carmustine, psoralen and ultraviolet A light (PUVA), and electron beam irradiation.[31–33] Patients who fail local skin therapy or develop advanced stage disease with visceral or lymph node involvement are candidates for systemic chemotherapy. Immune therapy with interferon alfa results in response rates up to 90% in untreated patients, but only 30% in heavily pretreated patients.[34] Combination chemotherapy is usually given to patients who have failed several trials of prior therapy and may result in 80% response rates, but responses are not durable and may range from 3 to 22 months.[35] Small studies have evaluated using purine analogues alone or in combination with other agents. In these heavily pretreated patients response rates ranged from 40 to 70% but responses were not durable.[36] Recently, retinoids have been shown to be effective in some patients with both early and advanced disease.[37]

Based on the response rates seen in CTCL patients in phase I trials, a multi-institutional phase III study to evaluate DAB$_{389}$IL-2 in CTCL patients was performed.[38] Patients had to have biopsy proven disease with 20% or greater lymphocytes within the biopsy stain positive for CD25 by immuno-histochemistry. Patients with stage I to III disease had to have disease recur after four or more prior treatments. Patients with stage IV disease had to have failed at least one previous therapy. Patients were randomly assigned to one of two dose levels of DAB$_{389}$IL-2 either 9 or 18 μg/kg/day dosed daily for 5 consecutive days and then repeated every 21 days. Patients were premedicated with acetaminophen and antihistamine 30 to 60 min prior to dosing. Of 276 patients screened, 71 met inclusion criteria. Objective responses were seen in 21 of 71 patients (30%), 8 of 35 patients on the 9 μg/kg/day arm, and 13 of 36 patients on the 18 μg/kg/day arm (p = NS). In patients with more advanced disease (stage IIb or greater) 2 of 21 patients (10%) responded in the 9 μg/kg/day arm compared to 9 of 24 patients (38%) in the 18 μg/kg/day arm (p = 0.07). Of the 50 patients classified as nonresponders, 25 patients were withdrawn early because of adverse events and were nonevaluable because predefined disease evaluation parameters were not met. However, 9 of these 25 patients had a 50% decrease in tumor burden at the time they discontinued from the study. Of the 23 patients classified with stable disease, 13 patients showed significant improvement in pruritus.

Treatment was discontinued early in 41 of 71 patients (58%) because of adverse events (37%), treatment failure (11%), and other reasons (10%). The first occurrence of adverse events in most patients was during the first treatment course (87%). Hypersensitivity events within 24 h of the infusion included dyspnea (20%), back pain (17%), hypotension (17%), chest tightness (13%), pruritus (13%), and flushing (13%). The above events were treated with stopping or slowing the infusion, and treatment with antihistamines and corticosteroids. The most frequent adverse event seen in 92% of patients was constitutional and gastrointestinal symptoms consisting of chills, fever, asthenia, nausea/vomiting, myalgia, arthralgia, headache, diarrhea, and anorexia. Infections of various types were reported in 56% of patients, but most infections were considered unrelated to treatment and typical of those seen in heavily pretreated patients with CTCL. Elevations of transaminases greater than five times the normal occurred in 17% of patients and hypo-albuminemia less than 2.3 g/dl occurred in 15% of patients. More than a 50% increase in creatinine occurred in 10% of patients and may have been related to VLS. VLS was seen in 25% of patients defined as edema, hypoalbumine-mia less than 2.9 g/dl, and/or hypotension occurring between days 1 to 14 of a treatment cycle. This syndrome was usually self-limited and treated by supportive measures. Grade 3 and 4 toxicities are summarized in Table 6.1. The authors reported that a serum albumin of 3.0 g/dl seemed to predict and may predispose for this syndrome. However, subsequent study of non-lymphoid malignancies and premedication with corticosteroids has resulted in a lower reported incidence of VLS.

Detectable levels of anti-DT antibodies by ELISA assay were seen in 19 of 60 patients (32%) and at the end of two courses of therapy this increased to 59 of 60 patients. Development of anti- $DAB_{389}IL$-2 neutralizing antibodies paralleled the increase in anti-$DAB_{389}IL$-2 ELISA titers and was not different in the two dosing groups or correlated to adverse events. Both responders and nonresponders had similar levels of antibody development, and devel-opment of antibodies to IL-2 did not preclude response. Pharmacokinetic analysis displayed two-compartment behavior with a distribution phase half-life of 2 to 5 min and a longer terminal phase with a half-life of 70 to 80 min. Clearance of $DAB_{389}IL$-2 increased by at least two-fold between course 1 and course 3, most likely because of the development of antibodies. This

Table 6.1 Percent grade 3 and 4 toxicities of DAB389IL-2 in CTCL patients

Toxicity	9 μg/kg/day	18 μg/kg/day
Constitutional symptoms	37	47
Hypotension	11	8
Pruritus	0	6
Vasodilation	3	3
Dyspnea	6	11
Chest pain	9	0
Gastrointestinal	20	36
Vascular leak	17	25
Transaminases > 5 × normal	14	24
Creatinine > 4 mg/dl	3	0
Neutropenia	3	3
Thrombocytopenia	0	3
Albumin less than 2.3 g/dl	14	17

Source: Adopted from E. Olsen, M. Duvic, A. Frankel, Y. Kim, et al., "Pivotal phase III trial of two dose levels of denileukin diftitox for the treatment of cutaneous T-cell lymphoma", *J Clin Oncol* 19, 2001, 376–388.

increased clearance may help explain why most patients who responded had responses within the first two treatment cycles and may be important in prolonged or repeated dosing.

In summary, DAB$_{389}$IL-2 has demonstrated acceptable toxicity and to be of clinical benefit to patients with advanced and refractory CTCL. DAB$_{389}$IL-2 is now FDA approved for the treatment of patients with CTCL whose tumor cells express CD25. Studies are now ongoing to evaluate DAB$_{389}$IL-2 in patients with low or absent levels of CD25 as well as continuing to study optimal dosing strategies.

DAB$_{389}$IL-2: what have we learned?

Decreasing DAB$_{389}$IL-2 toxicity

Patient safety and comfort is an important part of any therapy, and studies have been done to decrease the toxicities of DAB$_{389}$IL-2. Foss et al. retrospectively reviewed 15 patients with CTCL who received DAB$_{389}$IL-2 as dosed in the phase III study, however these patients were allowed to receive an oral loading dose of prednisone 20 mg prior to each DAB$_{389}$IL-2 infusion or dexamethasone 8 mg i.v. on day 1 followed by dexamethasone 8 mg i.v. prior to each dose. In addition, all patients received diphenhydramine and anti-emetics. Only three patients experienced acute infusion events and only two patients developed clinically apparent VLS. The authors concluded that steroid premedication significantly improves the tolerability of DAB$_{389}$IL-2.[39]

Two reports of DAB$_{389}$IL-2 in steroid resistant acute graft versus host disease (GVHD) reported a low incidence of VLS and infusion-related toxicities even though many of these patients had serum albumins below 3 g/dl.[40,41] This low incidence of VLS may be secondary to the concomitant use of high doses of corticosteroids or also to the lack of circulating malignant lymphoid cells.

Issues with detecting the IL-2R in the clinic

As previously discussed, the IL-2R is a heterotrimeric complex and the initial studies of DAB$_{389}$IL-2 required the presence of 20% or greater of p55 (CD25) on tumor specimens. However, detection of the intermediate or high affinity IL-2R may provide a better marker for determining patients to treat and follow responses. Although CD25 expression was a determinate for treatment with DAB$_{389}$IL-2, there was intrapatient variability within the same patient demonstrating different levels of CD25 expression from different biopsy specimens. Also, low or absent CD25 expression could be demonstrated because of receptor down modulation from DAB$_{389}$IL-2 therapy or possibly to selection of non-IL-2R expressing cells. Some patients with low or absent CD25 expression that had received prior DAB$_{389}$IL-2 and were retreated at relapse had responses. Studies are ongoing to determine the clinical benefit of $DAB_{389}IL$-2 in CTCL patients with low or absent levels of CD25 expression and to better characterize responses with markers for the intermediate-(CD122) and high-affinity (CD132) IL-2R. Other clinical issues with measuring the IL-2R are that the immunoassay for CD25 expression must be performed on frozen tissue, requiring rebiopsies, and any measure of the IL-2R will need to be standardized outside of reference laboratories.

Soluble IL-2R

It appears that the presence of soluble IL-2R (sIL-2R) has little impact on the effect of $DAB_{389}IL$-2. *In vitro* experiments demonstrated sIL-2R at levels <5000 pmol/l had no impact on the cytotoxicity of $DAB_{389}IL$-2, suggesting that sIL-2R competes poorly for this agent with the IL-2R.[4] In the phase I studies reported by LeMaistre et al. on both the 486 and 389 fusion toxins the presence of sIL-2R did not prevent antitumor response. However, in the phase I study of $DAB_{389}IL$-2 the CTCL patients who did not respond had significantly higher levels of sIL-2R than those patients who experienced an antitumor response. Future studies in CTCL patients will need to determine if a certain threshold of sIL-2R may inhibit response or whether sIL-2 is a marker of extensive tumor burden in this setting.

Issues with antibodies

Many people have received vaccinations for diphtheria and therefore may have antibodies against the active component of the fusion toxin. In the phase I study of $DAB_{389}IL$-2 40% of patients had detectable titers of anti-DT or anti-$DAB_{389}IL$-2 antibodies at study entry. After two cycles of therapy, antibody titers could be detected in almost all patients. In the phase I studies of both the 486 and 389 fusion toxins there was no apparent relationship between dose or dose level and peak antibody titer. Neither did the presence of antibodies preclude the opportunity for tumor response, suggesting that not all $DAB_{389}IL$-2 antibodies can effectively neutralize its activity. In the phase I study of $DAB_{389}IL$-2 there was a significant association of hypoalbuminemia with the presence of antibody titers to DT and $DAB_{389}IL$-2. In the phase III study clearance was increased between course 1 and 3 most likely because of antibody formation. Measuring antibody levels does not appear to be clinically necessary since steroid premedication can significantly diminish infusion-related reactions of $DAB_{389}IL$-2 and antibody levels do not correlate with response.

$DAB_{389}IL$-2 in combination with other agents

$DAB_{389}IL$-2 mediates its effect by binding to the high-affinity IL-2 receptor and upregulation of this receptor may enhance cellular intoxication by the fusion toxin. Rexinoids are ligands for transcription factors for nuclear receptors and are capable of upregulating high-affinity IL-2 receptor expression as well as enhancing susceptibility of cells to DAB$_{389}$IL-2.[42] A case report describes a CTCL patient with disease that had relapse after prior treatment with DAB$_{389}$IL-2 who then received the combination of DAB$_{389}$IL-2 18 μg/kg/day for 3 days repeated every 21 days with a retinoid, bexarotene, 225 mg/day. After completing 5 cycles of the combined therapy the patient obtained a CR and has maintained the response for over a year on maintenance bexarotene. The patient did develop a serum triglyceride level of 559 mg/dl and required treatment with atorvastatin as well as increased dosing of levothyroxine sodium tablets for preexisting hypothyroidism.[43] These investigators further discuss the results of a phase I trial with escalating doses of bexarotene combined with DAB$_{389}$IL-2 in patients with lymphoid malignancies. Bexarotene was administered to nine patients at 75, 150, 225, or 300 mg/day, and DAB$_{389}$IL-2 was given at 18 μg/kg/day daily for 3 days every 21 days. The bexarotene was given 7 days prior to the first dose of DAB$_{389}$IL-2. Six of nine patients responded with a CR in three patients. The six responding patients had upregulation of the CD25 subunit of the IL-2R in peripheral blood. The investigators concluded the combination of drugs was well tolerated with VLS noted in three patients, grade 1/2 hepatotoxicity in five patients, and grade 3/4 lymphopenia in five patients. The most common side effects were hypothyroidism in 67% of patients and hypertriglyceridemia in 87% of patients.[44]

DAB$_{389}$IL-2 in the treatment of other lymphoid malignancies

As mentioned in the previous section, a small number of patients with B-cell NHL had responded to DAB$_{389}$IL-2 in the phase I/II studies. Preliminary data on patients with fludarabine refractory CLL treated with DAB$_{389}$IL-2 has been reported. Eighteen patients were treated with DAB$_{389}$IL-2 at 9 or 18 μg/kg/day as a 1 h infusion for 5 days given every 21 days. Ten patients had received at least three courses, of which nine patients had reductions in peripheral circulating malignant cells. Six of the ten patients also had some reduction of peripheral lymphadenopathy. Bone marrow responses showed six of the ten patients had 50% or greater reduction in CLL marrow index. Overall, one of twelve evaluable patients had a partial response. Toxicities were similar to previous reports and patients in this study were not given steroid premedication.[45] In another study 31 patients with relapsed or refractory B- and T-cell NHL were given DAB$_{389}$IL-2 at 18 μg/kg/day for 5 days over 60 min and then repeated every 3 weeks to a maximum of 8 cycles. Of 25 patients evaluable for response, one patient with follicular mixed NHL had a CR and four other patients had PRs. Two of the patients that had a PR also had cells that were negative for CD25. Patients were allowed premedication with steroids to prevent infusion-related reactions and VLS and overall the investigators found the drug to be well tolerated.[46] Case reports have also demonstrated activity in primary refractory T-cell NHL[47] and T-cell anaplastic large cell lymphoma.[48]

The recommended treatment regimen for CTCL is 9 or 18 mcg/kg/day intravenously for 5 consecutive days repeated every 21 days based on the dosing used in the pivotal phase III study. There was no demonstrated dose–response effect and no significant difference in toxicity between the two doses. However, there was a trend toward a higher response rate in patients with more advanced disease in the higher dose group, as well as increased hypotension and hypoalbuminemia. It is possible that the treatment of more patients with advanced disease at the higher dose level with corticosteroid premedication may be able to detect a significant difference in disease response. There are ongoing phase IV studies comparing these same dosing schedules in patients with earlier stage disease and less prior therapy. Dose-limiting toxicities occurred at a dose of 31 μg/kg/day and higher doses have not been evaluated in humans. Prolonged or continuous infusions would probably not be beneficial with $DAB_{389}IL$-2 because the half-life of the fusion toxin allows adequate presentation of the toxin to susceptible cells. The optimal schedule and length of treatment has not been determined, and other phase II studies may further delineate the best schedule based on time to response, duration of response, and cost.

Combination with other chemotherapy

Chemotherapy drug resistance is often a mode of therapy failure and ultimately disease relapse in many different neoplastic diseases. Drug resistance is often mediated by transmembrane transporter molecules, including P-glycoprotein or the multi-drug resistance associated protein, or associated with the inactivation of the p53 tumor suppressor gene, as well as overexpression of the anti-apoptotic protein bcl-2. Fusion toxins may circumvent these modes of drug resistance, as exemplified by Perentesis et al. who used a fusion of DT with granulocyte-macrophage colony stimulating factor (GMCSF) to kill greater than 99% of primary leukemic progenitor cells from therapy refractory AML patients.[49] Targeting the high affinity IL-2R with $DAB_{389}IL$-2 in addition to cytotoxic chemotherapy may use multiple pathways to destroy malignant clones and therefore prevent drug resistance.

Immune-mediated diseases

Immune-mediated nonmalignant diseases may be treated by $DAB_{389}IL$-2 because of its ability to target activated T-cells yet spare the hematologic and nonhematologic toxicity of many other immunosuppressive agents. Some patients with long-standing rheumatoid arthritis despite the use of steroids, NSAIDS, or methotrexate showed a greater than 25% improvement in both tender and swollen joints counts when treated with $DAB_{389}IL$-2.[50] In several small trials using both $DAB_{486}IL$-2 and $DAB_{389}IL$-2 up to 33% of patients had some response to treatment.[51] In another study of psoriasis patients 30 of 91 patients (33%) who received $DAB_{389}IL$-2 had a 50% decrease in disease activity.[52,53] In the aforementioned studies patients experienced similar toxicities as those mentioned for lymphoma patients, however, the frequency and severity of events was generally lower. The doses and schedules of DAB$_{389}$IL-2 were done in a variety of ways and no clear dose response or optimal schedule can be deduced.

GVHD occurs after allogeneic hematopoietic stem cell transplantation (ASCT) when donor T-cells recognize host antigen as nonself and mediate tissue damage. GVHD is now one of the leading causes of morbidity and mortality after ASCT and can be associated with a greater than 80% 1 year mortality if refractory to treatment with corticosteroids. An initial experience with DAB$_{389}$IL-2 in steroid refractory GVHD described 11 patients treated with a dose of 4.5 µg/kg/day for 5 days and then the same dose repeated weekly for 4 weeks.[40] Ten patients experienced a greater than 1 grade response, with three patients later progressing before the end of the study. Five patients demonstrated a CR and two patients had a PR but later died of infection before the end of treatment. Overall survival was 55% (6/10) with two patients in continued CR and on tapering doses of steroids. Two patients had aGVHD flares after the end of treatment and two patients had progressed while on treatment and were placed on other therapy. Infusion related toxicities were minimal, and only one patient experienced VLS. All patients were on high doses of corticosteroids, which may have helped ameliorate the infusion-related toxicities but also contributed to the infectious complications of these patients. One infection with nocardia and aspergillus each and two cases of CMV infection were seen in study patients. Only one patient relapsed with their underlying hematologic malignancy after treatment with DAB$_{389}$IL-2. This patient had a refractory intermediate grade NHL and relapsed their NHL with ongoing steroid refractory aGVHD a month after the DAB$_{389}$IL-2 was stopped for VLS. In another report 14 patients were treated with DAB$_{389}$IL-2 for steroid refractory GVHD who had failed other immunosuppressive agents.[41] Patients were treated at two dose levels, seven patients received 9 µg/kg/day on days 1 and 15, and seven patients received the same dose on days 1, 3, 5, 15, 17, and 19. Four patients obtained a CR, four patients a PR, and three patients did not respond. All patients who achieved a CR were on regimen 2. One patient experienced a mild infusion-related reaction and two patients had transaminitis. Overall, DAB$_{389}$IL-2 was well tolerated in this population of patients and demonstrated activity in treating steroid-resistant GVHD. It is possible that the high dose of steroids or the lack of active malignancy in these patients contributed to the paucity of infusion-related events.

Summary

DAB$_{389}$IL-2 is a fusion toxin that targets the high-affinity IL-2R and efficiently intoxicates and kills cells bearing this receptor. This selective action has proven clinically beneficial to patients with CTCL that have failed other therapies, and further study in patients with earlier stage disease, who may or may not express the IL-2R, may further widen its therapeutic application. Optimal dosing and schedules of delivery of DAB$_{389}$IL-2 in CTCL, as well as other malignancies and immune-mediated diseases, are currently under study. Combining DAB$_{389}$IL-2 with drugs that potentiate IL-2R expression or in combination with other conventional cytotoxic drugs may help increase response rates and prevent or overcome drug resistance.

References

1. M. Eapen, M. Horowitz, "Report on state of the art in blood and marrow transplantation — the IBMTR/ABMTR summary slides," *IBMTR/ABMTR Newsletter* 9, 2002, 4–11.
2. B.L. Kagan, A. Finkelstein, M. Colombini, "Diphtheria toxin fragment forms large pores in phospholipid bilayer membranes," *Proc Natl Acad Sci USA* 78, 1981, 4950–4954.
3. R. Fuchs, S. Schmid, I. Mellman, "A possible role for Na+, K+-ATPase in regulating ATP-dependent endosome acidification," *Proc Natl Acad Sci USA* 86, 1998, 539–543.
4. C.F. LeMaistre, M. Saleh, T. Kuzel, F. Foss, et al. "Phase I trial of a ligand fusion-protein (DAB389IL-2) in lymphomas expressing the receptor for intereukin-2," *Blood* 91, 1998, 399–405.
5. F. Craig, P. Banks, "Detection of the alpha and beta components of the interleukin-2 receptor using immunologic techniques," *Mol Pathol* 5, 1992, 118a (Abstract).
6. R.J. Robb, W.C. Greene, "Internalization of interleukin-2 is mediated by the gamma chain of the high affinity interleukin-2 receptor," *J Exp Med* 165, 1987, 1201–1206.
7. T.A. Waldmann, "The structure, function, and expression of interleukin-2 receptors on normal and malignant lymphocytes," *Science* 232, 1986, 727–732.
8. T.A. Waldmann, "The multichain interleukin 2 receptor. A target for immunotherapy in lymphoma, autoimmune disorders, and organ allografts," *JAMA* 263, 1990, 272–274.
9. T. Taniguchi, Y. Minami, "The IL-2/IL-2 receptor system: a current overview," *Cell* 73, 1993, 5–8.
10. J.A. Strauchen, B.A. Breakstone, "IL-2 receptor expression in human lymphoid lesions. Immunohistochemical study of 166 cases," *Am J Pathol* 126, 1987, 506–512.
11. K. Sheibani, C.D. Winberg, S. van de Velde, et al., "Distribution of lymphocytes with interleukin-2 receptors (TAC antigens) in reactive lymphoproliferative processes, Hodgkin's disease, and non-Hodgkin's lymphomas. An immunohistologic study of 300 cases," *Am J Pathol* 127, 1987, 27–37.
12. D. Barnett, G.A. Wilson, A.C. Lawrence, et al., "The interleukin-2 receptor and its expression in the acute leukaemias and lymphoproliferative disorders," *Dis Markers* 6, 1988, 133–139.
13. A. Rosolen, M. Nakanishi, D.G. Poplack, et al., "Expression of interleukin-2 receptor beta subunit in hematopoietic malignancies," *Blood* 73, 1989, 1968–1972.
14. D. Williams, K. Parker, P. Bishai, et al., "Diphtheria toxin receptor binding domain substitution with interleukin-2: genetic construction and properties of a diphtheria toxin-related interleukin-2 fusion protein," *Protein Eng* 1, 1987, 493.
15. P. Bacha, D. Williams, C. Waters, J. Murphy, T. Strom, "Interleukin-2 receptor cytotoxicity: Interleukin-2 receptor-mediated action of a diphtheria toxin-related interleukin-2 fusion protein," *J Exp Med* 167, 1988, 612.

16. G. Walz, B. Zanker, K. Brand, et al., "Sequential effects of interleukin 2-diphtheria toxin fusion protein on T-cell activation," *Proc Natl Acad Sci USA* 86, 1989, 9485–9488.
17. P. Bacha, D.P. Williams, C. Waters, J.R. Murphy, T.B. Strom, "Interleukin-2 receptor cytotoxicity: Interleukin-2 receptor-mediated action of a diphtheria toxin-related interleukin-2 fusion protein," *J Exp Med* 167, 1988, 612.
18. C.A. Waters, P.A. Schimke, C.E. Snider, T.B. Strom, J.R. Murphy, "Receptor binding requirements for entry of a diphtheria toxin-related interleukin 2 fusion protein into cells," *Eur J Immunol* 20, 1990, 785.
19. C.F. LeMaistre, C. Meneghetti, M. Rosenblum, J. Reuben, et al., "Phase I trial of an interleukin-2 (IL-2) fusion toxin (DAB$_{486}$IL-2) in hematologic malignancies expressing the IL-2 receptor," *Blood* 79, 1992, 2547–2554.
20. C.F. LeMaistre, F. Craig, C. Meneghetti, B. McMullin, et al., "Phase I trial of a 90 minute infusion of the fusion toxin DAB$_{486}$IL-2 in hematologic cancers," *Cancer Res* 53, 1993, 3930–3934.
21. P. Hesketh, P. Caguioa, H. Koa, H. Dewey, et al., "Clinical activity of a cytotoxic fusion protein in the treatment of cutaneous T-cell lymphoma," *J Clin Oncol* 11, 1993, 1682–1690.
22. F. Foss, T. Borkowski, M. Gilliom, et al. "Chimeric fusion protein toxin DAB$_{486}$IL-2 in advanced mycosis fungoides and the Sezary syndrome: correlation of activity and interleukin-2 receptor expression in a phase II study," *Blood* 84, 1994, 1765–1774.
23. C.F. LeMaistre, M. Saleh, T. Kuzel, F. Foss, et al., "Phase I trial of a ligand fusion-protein (DAB389IL-2) in lymphomas expressing the receptor for interleukin-2," *Blood* 91, 1998, 399–405.
24. Ontak package insert, Ligand Pharmaceuticals Inc, San Diego CA, 1999.
25. P.A. Bacha, S.E. Forte, D.M. McCarthy, "Impact of interleukin-2 receptor-targeted cytotoxins on a unique model of murine interleukin-2 receptor-expressing malignancy," *Int J Cancer* 49, 1991, 96–101.
26. C.F. LeMaistre, M. Saleh, T. Kuzel, F. Foss, et al., "Phase I trial of a ligand fusion-protein (DAB$_{389}$IL-2) in lymphomas expressing the receptor for interleukin-2," *Blood* 91, 1998, 399–405.
27. M. Saleh, C.F. LeMaistre, T. Kuzel, F. Foss, et al., "Antitumor activity of DAB$_{389}$IL-2 fusion toxin in mycosis fungoides," *J Am Acad Dermatol* 39, 1998, 63–73.
28. R. Hoppe, G. Wood, E. Abel, "Mycosis fungoides and the Sezary syndrome: Pathology, staging and treatment," *Curr Prob Cancer* 14, 1990, 297–361.
29. G.S. Wood, L.M. Weiss, R.A. Warnke, et al., "The immunopathology of cutaneous lymphomas: Immunophenotypic and immunogenotypic characteristics," *Semin Dermatol* 5, 1986, 334–345.
30. P.A. Bunn, S.I. Lambert, "Report of the committee on staging and classification of CTCL," *Cancer Treat Rep* 63, 1979, 726–727.
31. H.S. Zackheim, E.H. Epstein, W.R. Crain, "Topical camustine (BCNU) for cutaneous T-cell lymphoma: a 15 year experience in 143 patients," *J Am Acad Dermatol* 22, 1990, 802–810.
32. E.C. Vonderheid, E.T. Tan, A.F. Kantor, et al., "Long term efficacy, curative potential, and carcinogenecity of topical mechlorethamine chemotherapy in cutaneous T-cell lymphoma," *J Am Acad Dermatol* 20, 1989, 416–428.
33. F.J. Kaye, P.A. Bunn, S.M. Steinberg, et al., "A randomized trial comparing combination electron beam radiation and chemotherapy with topical therapy in the initial treatment of mycosis fungoides," *N Engl J Med* 321, 1989, 1784–1790.

34. E.A. Olsen, S. Rosen, R. Vollmer, et al., "Interferon alfa 2a in the treatment of cutaneous T-cell lymphoma," *J Am Acad Dermatol* 20, 1989, 395–407.

35. G. Akpek, H. Koh, S. Bogen, et al., "Chemotherapy with etoposide, vincristine, doxorubicin, bolus cyclophosphamide, and oral prednisone in patients with refractory cutaneous T-cell lymphoma," *Cancer* 86, 1999, 1368–1376.

36. D. Greiner, E. Olsen, G. Petroni, "Pentostatin in the treatment of cutaneous T-cell lymphoma," *J Am Acad Dermatol* 36, 1997, 905–955.

37. M.A. Smith, D.R. Parkinson, B.D. Cheson, et al., "Retinoids in cancer therapy," *J Clin Oncol* 10, 1992, 839–864.

38. E. Olsen, M. Duvic, A. Frankel, Y. Kim, et al. "Pivotal phase III trial of two dose levels of denileukin diftitox for the treatment of cutaneous T-cell lymphoma," *J Clin Oncol* 19, 2001, 376–388.

39. F. Foss, P. Bacha, K. Osann, M. Demierre, T. Bell, T. Kuzel, "Biological correlates of acute hypersensitivity events with DAB$_{389}$IL-2 (Denileukin Diftitox, Ontak) in cutaneous T-cell lymphoma: decreased frequency and severity with steroid premedication," *Clin Lymphoma* 1, 2001, 298–302.

40. P. Shaughnessy, C. Bachier, M. Grimley, C. Freytes, C.F. LeMaistre, "Phase II study of denileukin diftitox (Ontak) in the treatment of steroid resistant acute graft versus host disease," *Biol Blood Marrow Transpl* 9, 2003, abst. 110.

41. V. Ho, D. Zahrieh, D. Neuberg, M. Munoz, et al., "Safety and efficacy of denileukin diftitox (Ontak) in patients with steroid refractory graft-versus host disease (GVHD) after allogeneic hematopoietic stem cell transplantation (HSCT)," *Blood* 100, 2002, abst. 1638.

42. G. Gorgun, F. Foss, "Immunomodulatory effects of RXR rexinoids: modulation of high-affinity IL-2R expression enhances susceptibility to denileukin diftitox," *Blood* 100, 2002, 1399–1403.

43. G. DiVenuti, F. Foss, "Cutaneous T-cell lymphoma treated with a combination of denileukin diftitox and bexarotene," *Case Studies in Lymphoma* 1, 2002, 2–7.

44. G. Divenuti, F. Foss, "Phase I dose escalation study of Targretin and Ontak in hematologic malignancies: upregulation of IL2R expression by low dose Targretin," *Blood* 98, 2001, abst. 2520.

45. A.E. Frankel, D.R. Fleming, R. Gartenhaus, B.L. Powell, J.H. Black, C.F. Leftwich, "Diphtheria fusion protein Ontak therapy of patients with fludarabine-refractory chronic lymphocytic leukemia," *Blood* 100, 2002, abst. 5020.

46. N. Dang, F. Hagemeister, L. Fayad, B. Pro, et al. "Phase II study of delineukin diftitox (Ontak) for relapsed refractory B and T-cell non Hodgkin's lymphoma," *Blood* 100, 2002, abst. 1405.

47. J. Fleagle, "Primary refractory T-cell NHL treated with denileukin diftitox (Ontak)," *Case Studies in Oncology* 3, 2001, 3–7.

48. Meha et al. "Treatment of T-cell anaplastic large cell lymphoma (ALCL) relapsing after autologous stem cell transplant (ASCT) with denileukin diftitox (Ontak)," *Blood* 100, 2002, abst. 4772.

49. J.P. Perentesis, A.E. Bendel, Y. Shao, B. Warman, et al. "Granulocyte-macrophage colony-stimulating factor receptor-targeted therapy of chemotherapy- and radiation-resistant hyman myeloid leukemias," *Leuk Lymphoma* 25, 1997, 247–56.

50. K. Sewell, L. Moreland, D. Furst, et al. "Phase II: open label trial of DAB$_{389}$IL-2 administered up to four times a year to patients with active rheumatoid arthritis," *Arthritis Rheum* 37, 1994, S341.

51. J. Kremer, G. Petrillo, W. Rigby, "Phase I/II open label trial of DAB$_{389}$IL-2 administered to patients with active rheumatoid arthritis receiving treatment with methotrexate," *Arthritis Rheum* 37, 1994, S341.
52. A. Gottlieb, P. Bacha, K. Parker, V. Strand, "Use of interleukin-2 fusion protein, DAB$_{389}$IL-2, for the treatment of psoriasis," *Dermatologic Ther* 5, 1998, 48–63.
53. S. Gottlieb, P. Gilleaudeau, R. Johnson, L. Estes, et al., "Response of psoriasis to a lymphocyte selective toxin (DAB$_{389}$IL-2) suggests primary immune, but not keratinocyte, pathogenic basis," *Nature Med* 1, 1995, 442–447.

chapter seven

Transferrin diphtheria toxin for brain tumor therapy

Simon Long and Patrick Rossi

Contents

Introduction

Primary brain tumors are a significant cause of morbidity and mortality with new malignant brain tumors diagnosed in approximately 17,000 Americans each year (American Cancer Society, 2002). The most common primary brain tumors are glial tumors, including the highly malignant glioblastoma multiforme (GBM). Malignant gliomas constitute at least 35% of all primary brain tumors and are the third leading cause of death from cancer in persons 15 to 34 years of age (Salcman 1990: 95). In addition, recent evidence suggests that the incidence of primary brain tumors is increasing among the elderly (Grieg et al. 1990: 1621).

The prognosis for patients with malignant gliomas is poor and despite aggressive therapy, including surgery, postoperative high-dose radiation and chemotherapy, the median survival from diagnosis in patients with GBM is less than 1 year (Schold et al. 1997: 51). Standard chemotherapy has only had a modest effect on the survival of patients with high-grade glioma. A recently published meta-analysis of 3004 patients in 12 randomized trials of chemotherapy indicated an absolute increase in 1-year survival of 6% (Glioma Meta-analysis Trialists Group 2002: 1011). Recent developments in the basic cellular biology of gliomas is now leading to a better understanding of these tumors, which may ultimately lead to improved therapies and outcomes. Furthermore, improved drug delivery technology is enabling the exposure of tumor cells to increased amounts of therapeutic agents whilst limiting systemic toxicity.

Targeted protein toxins

The poor prognosis for central nervous system (CNS) malignancy is related to a number of factors including poor distribution of systemically given drugs to tumor tissue due to the blood-brain barrier (BBB) or poor diffusion within the tumor for locally administered therapies (e.g., carmustine impregnated polymer implants), and the lack of potent agents with adequate tumor specificity. Targeting to specific cell receptors provides the possibility of creating novel therapeutic agents with greater tumor specificity than conventional chemotherapy. Targeted protein toxins consist of a targeting polypeptide linked to a peptide toxin. Monoclonal antibodies against tumor-associated antigens and other binding moieties, which provide tumor selectivity, have been conjugated with radionuclides and with various toxins (Fitzgerald and Pastan 1989: 1455; Frankel et al. 2000: 326; Gilliland et al. 1980: 4539; Jansen et al. 1982: 185; Lashford et al. 1988: 857; Trowbridge and Domingo 1981: 171; Uhr 1984: i; Youle and Colombatti 1986: 173).

Investigators have studied a targeted protein toxin, which uses the physiological binding of human transferrin (Tf) to transferrin receptors (TfR) expressed on metabolically active cells, to achieve tumor specificity. This targeted protein toxin is transferrin-CRM107 (Tf-CRM107, TransMID™), a conjugate of human Tf and diphtheria toxin (DT) with a point mutation (CRM107) that inactivates the nonspecific binding to mammalian cells (Greenfield et al. 1987: 536). DT belongs to a group of protein toxins that consist of A and B subunits. The B subunit is responsible for toxin-binding to the cell surface and for translocation of the A-chain into the cytosol, which catalyzes the transfer of adenosine diphosphateribose to elongation factor 2 (EF-2). This prevents the transfer of peptidyl-tRNA on ribosomes, thereby inhibiting protein synthesis and killing the cell. The modification of CRM107 consists of one amino acid change in the B-chain (phenylalanine for serine at position 525) that reduces binding 8000-fold, but leaves translocation and enzymatic functions intact. The modified DT (CRM107) and Tf are joined by a stable, nonreducible thioether bond (Johnson et al. 1989: 240), and the resulting conjugate has a molecular weight of approximately 140 kD.

TfRs transport iron into cells and are overexpressed on rapidly dividing cells, most notably on hematopoietic cells and various tumor cells, including glioblastoma cells. This is thought to reflect the increased iron requirements of rapidly dividing cells (Faulk et al. 1980: 390; Galbraith et al. 1980: 215; Gatter et al. 1983: 539; Larrick and Cresswell 1979: 579; Shindelman et al. 1981: 329; Trowbridge et al. 1984: 925; Hamilton et al. 1984: 2285; Klausner et al. 1983: 4715). Further studies have demonstrated an even higher expression on glioblastoma and medulloblastoma tumor cell lines in comparison to the erythroleukemia cell line K562, which is known to express TfR at a high level of approximately 1.6×10^5 TfR sites/cell (Zovickian 1987: 850). In contrast to this, TfRs in normal brain tissue are sparse and are largely restricted to the luminal surface of brain capillaries (Angelova-Gateva 1980: 27; Connor and Fine 1986: 319; Hill et al. 1985: 4553; Jefferies et al. 1984: 162).

Preclinical studies

Several *in vitro* cytotoxicity assays have been performed to evaluate the potency of Tf-CRM107.

The cytotoxicity and specificity of Tf-CRM107 in comparison to free toxin CRM107 was investigated using a protein synthesis inhibition assay in three different cell lines (Jurkat, K562, and SNB75). In addition, the inhibition of toxicity by free transferrin was evaluated. The IC_{50} values ranged between 2.0×10^{-12} and 1.7×10^{-11} M for Tf-CRM107 and between 2.5×10^{-8} and 5.4×10^{-7} M for the free toxin CRM107 in the three cell lines tested. This indicated that the free toxin was 1,000- to 100,000-fold less toxic than the conjugated toxin. A dose of 3.6×10^{-10} M of Tf-CRM107 decreased protein synthesis in K562 cells to 0.7% of the control. This effect could be completely inhibited by adding 300 µg/ml of free transferrin demonstrating that the cytotoxic effect of Tf-CRM107 is specifically mediated by the binding of Tf-CRM107 to the transferrin receptor (Johnson 1988: 1295).

The *in vitro* efficacy of Tf-CRM107 compared to the free unconjugated toxin CRM107 was also evaluated on human tumor-derived cell lines. Four medulloblastoma (two established cell lines and two primary medulloblastoma cultures), three glioblastoma, and three breast carcinoma cell lines were examined using an inhibition of protein synthesis assay. The IC_{50} values ranged from 3.9×10^{-13} to 1.1×10^{-10} M. The free toxin CRM107 was 10,000- to 1,000,000-fold less toxic than the conjugated toxin (Johnson et al. 1989: 240)

In a further study the *in vitro* efficacy and cytotoxicity of Tf-CRM107 was investigated in three other cell lines (bladder cancer, glioblastoma, and medulloblastoma) and on operative specimens from three pediatric brain tumors (two pilocytic astrocytomas and one craniopharyngioma). The IC_{50} values ranged between 1.6×10^{-11} and 4.0×10^{-10} M for the tumor cell lines and between 2.0×10^{-11} and $>10^5$ M (craniopharyngioma) for the primary tumor cells.

The *in vivo* efficacy of Tf-CRM107 was shown in a direct intratumoral injection study in nude mice bearing solid human gliomas (U251). All three doses of Tf-CRM107 (0.1, 1.0, and 10 µg) resulted in significant dose-dependent inhibition of tumor growth and tumor weight reduction. Treatment with unconjugated CRM107 was 10 to 100 times less effective in growth inhibition. Animals treated with phosphate buffered saline (PBS) showed continuing tumor growth (Laske 1994: 520)

Toxicology

In a single dose, intracerebral, stereotactic infusion study with Tf-CRM107 in rats (Oldfield 1992) the maximum tolerated dose (MTD), defined as the highest dose that resulted in no animal deaths, was 0.33 µg/kg Tf-CRM107 when infusing a volume of 10 µl, and a dose of 1.67 µg/kg Tf-CRM107 when infusing a volume of 50 µl, based on an infusate concentration of 7.0×10^{-8} M. When infusing volumes of 10 µl or 50 µl, infusate concentrations of 3.5×10^{-8} or 7.0×10^{-8} M, respectively, were the lowest concentrations at which histologic changes occurred. These changes were characterized by encephalomalacia of the right frontal lobe.

In single-dose, intrathecal injection studies with Tf-CRM107, the MTD consistent with survival corresponded to a cerebrospinal fluid (CSF) concentration of 2.0×10^{-9} M in guinea pigs and of more than 2.0×10^{-9} M in rhesus monkeys (Johnson 1989: 240).

The whole body distribution of Tf-CRM107 was studied following single bolus intra-cerebroventricular (i.c.v.) and intravenous (i.v.) injections into healthy mice and rats. Tf-CRM107 escapes from the brain in the absence of a tumor and is mainly distributed in the lungs, liver, kidneys, heart, and gastrointestinal tract. Tf-CRM107 was detected in plasma but not urine following intravenous administration to rats with a mean plasma half-life of 8.7 and 18.1 h after doses of 1 and 10 µg/kg, respectively. Tf-CRM107 did not appear to be metabolized either *in vitro* in homogenates or microsomes or *in vivo* in rats.

Drug delivery

A major hurdle in the treatment of brain tumors is delivering enough of the therapeutic agent to the site of the tumor and surrounding infiltrated tissue whilst avoiding neurotoxicity. Drug administration to the CNS is impaired because of the BBB, which, when not disrupted, will prevent the passage of molecules that are larger than 180 kD into the brain. One strategy that has been used to circumvent the BBB is direct intratumoral chemotherapy. This has been administered by direct injection, intracavity instillation, intracavity topical application, chronic low flow microinfusion, and by controlled release from polymer implants. However, the efficacy of direct intratumoral chemotherapy is restricted by the poor diffusion of drug through the tumor relative to tissue clearance so that only a small volume of tissue surrounding the drug source is treated. The technique of convection-enhanced delivery (CED) uses a positive pressure gradient to enhance the distribution of both small and large molecules within the brain, including high molecular weight proteins (Bobo et al. 1994: 2076).

The biodistribution of intracerebrally infused ^{111}In-transferrin was studied in 12 anesthetized mongrel cats (Bobo et al. 1994: 2076). Three volumes of mock cerebrospinal fluid containing ^{111}In-transferrin were infused into the corona radiata bilaterally over 1 to 4 h. All the interstitial infusions were well tolerated and were not associated with any hemodynamic instability during the infusion. A linear increase of the intracerebral volume of distribution with the infusion volume was observed, and the systemic concentrations of ^{111}In-transferrin were at least four orders of magnitude lower than the intracerebral concentrations. These data indicated that interstitial infusion can achieve a high concentration of compound locally with low systemic concentrations (<1% of brain concentrations).

Phase I clinical study

A phase I clinical study of regional therapy in patients with refractory malignant brain tumors was performed with Tf-CRM107 at the National Institute of Health (NIH) under a physician-sponsored IND (Study PHCN-001).

The objectives of the study were to determine the toxicity of Tf-CRM107 when delivered to malignant brain tumors by intratumoral and peritumoral slow infusion in a dose escalation schedule and to determine the antitumor effect of such regional therapy.

Patients enrolled in the trial had a malignant brain tumor (primary or metastatic) that had failed standard therapy (including surgery and radiation therapy) and was radiographically documented as recurrent or progressive. Patients were over 18 years of age with a Karnofsky Performance Scale Score >30 and had no other cancer therapy within 4 weeks of inclusion into the trial. Other exclusion criteria included: pregnancy, active infection, abnormal liver or renal function, thrombocytopenia, neutropenia, abnormal clotting function, or infection with human immunodeficiency virus (HIV).

A total of 28 patients (32 tumors), 13 women and 15 men, with an average age of 46.5 years were included in the study. All had previously been treated with surgery and radiotherapy for their tumors, and 20 patients had previously received chemotherapy. The median Karnofsky Performance Scale score at baseline was 80, with a range of 30 to 100. The tumor location was the left side in 14 patients, right side in 11 and bilateral in 3. Of these tumors, eight were fronto-parietal, seven were frontal, seven were temporal, two were occipito-parietal, one was occipital, one was parietal, one was fronto-temporal and one was tempo-parietal.

The histopathologic diagnoses were 16 glioblastoma, 8 anaplastic astrocytoma, 2 metastatic disease (lung carcinoma), 1 anaplastic oligodendroglioma, and 1 anaplastic glial tumor. Of 24 patients tested 19 had tumors that stained positive for transferrin receptors (i.e. ≥10 cells). The median tumor volume was 10.6 cm^3, with a range of 0.6 to 50 cm^3 (baseline tumor volume measurements were made for only 28 of the 32 tumors). Anti-DT titers were below the lower limit of normal in 17 of the 28 patients prior to the infusion of Tf-CRM107.

Tf-CRM107 was administered via CED through CT-guided, stereotactically placed, silastic infusion catheters placed directly into the tumor or peritumoral region. The infusions were initiated at a slow rate and increased in a step-wise manner. The rates for each of the complete infusions ranged from 0.021 to 0.714 ml/h, administered through one to three catheters.

The initial Tf-CRM107 total dose was 0.5 μg, in a volume of 5 ml (i.e., a concentration of 0.1 μg/ml). This dose was escalated to a maximum dose of 128 μg and a maximum volume of 240 ml (i.e., a maximum concentration of 3.2 μg/ml). Patients were treated with intravenous prophylactic antibiotics and steroids prior to and for the duration of the infusion of Tf-CRM107. Antibiotics were stopped at the end of the infusion, but steroid doses were tapered as clinically indicated.

Magnetic resonance imaging (MRI), with and without gadolinium enhancement, was performed at baseline, immediately following each infusion. During and immediately after the Tf-CRM107 infusion, and at subsequent follow-up visits, neurological status, serum chemistries, anti-DT toxin antibody titers, blood count, and anticoagulation profiles were monitored. Patients returned every 4 to 6 weeks for evaluation (and re-treatment, if necessary) for the first 6 months following the last dose of Tf-CRM107 and patients were followed up until death.

The criteria for tumor response in this trial were as follows:

- *Complete Responder (CR)*: No remaining tumor on MRI at any measured time point.
- *Partial Responder (PR)*: Greater than 50% decrease in tumor volume by MRI at any measured time point.
- *Nonresponder (NR)*: No observable decrease in the size of the tumor by MRI scan or a progression (increase) in the size of the tumor.

Clinical chemistry and hematology values were graded according to the National Cancer Institute (NCI) toxicity grading scale.

Twenty-eight patients were administered a total of 65 infusions with Tf-CRM107. On a per patient basis, the median total number of catheters placed for each patient was 4 (range 1 to 10). The median total number of days of infusion was 19.5 (range 3 to 45). The median total volume of drug infused was 165 ml (range 5.5 to 393 ml) and the median total dose given was 68.5 μg (range 0.55 to 233.6 μg).

In total, 28 patients with 32 separate tumors were evaluated. Of these 32 tumors, two complete responses were reported. In addition, there were partial responses reported for eight tumors, nineteen were reported as nonresponders, and three were not evaluable. Two patients with metastatic lung carcinoma underwent tumor resection after therapy and were therefore not evaluable for response and one patient withdrew from the trial after 6 weeks and could not be evaluated for radiographic response. Of the responders, four of the ten were on larger steroid doses at the time the response was apparent compared to baseline. Four patients responded after receiving one infusion, and four patients responded after receiving two infusions. Of the remaining two responders, one responded after three infusions, and the other responded after five infusions. TfR expression did not appear to correlate with tumor response. As of the final follow-up date the median patient survival/duration of follow-up was 273 days longer (451 versus 178) in the responders versus nonresponders.

The 65 infusions administered to these 28 patients included seven drug concentrations (0.1 to 3.2 μg/ml) and 20 dose volumes (0.5 to 240.0 ml) resulting in 29 different doses (0.05 to 128.0 μg). The number of infusions administered ranged from 1 to 6, which were separated by periods of 25 to 287 days (mean 81.4 days). The mean time between the first and second infusion was 49 (25 to 134) days. The mean dose was 35.5 μg and the median dose was 26.8 μg. Five infusions of 26.8 μg (40 ml of 0.67 μg/ml) were administered.

Twenty of the 28 patients (71%) reported at least one adverse event (AE). A total of 52 AEs were reported by these 20 patients. Of the 52 AEs 18 were judged by the investigator to be unrelated to the treatment. These events were attributed directly to the patients' underlying disease, concomitant medication, or procedures.

Neurologic events were the most frequently reported AEs, accounting for 41 of the 52 events (79%). Of 31 neurological AEs 20 reported for the first 21 patients treated were considered serious (this classification is not available for AEs reported for the final seven patients treated). Of these, only four (dysphasia, right-sided weakness affecting the arm and two episodes of left arm weakness) were considered to be related to Tf-CRM107 and six were of uncertain relationship.

Of the 52 AEs, 39 (75%) occurred within 50 days of the start of infusion. Six AEs occurred more than 100 days (111 to 522 days) after the start of infusion (five occurrences of neurological toxicity and one occurrence of hydrocephalus). More than half (27/52; 52%) of the AEs occurred within 17 days after the start of infusion; the majority of these (24/27; 89%) were neurologic adverse events.

All 28 patients had at least one abnormality in hematology or serum chemistry parameters prior to receiving Tf-CRM107. However, no laboratory parameters other than ALT, AST, and albumin showed increasing or decreasing trends during the study. The trends that were observed in ALT (increasing), AST (increasing), and albumin (decreasing) were transient and not severe. There were no clinical signs associated with the changes. The relationship of these trends to Tf-CRM107 administration is complicated by anti-epileptic medication that the majority of the patients received throughout the trial.

Sustained neurotoxicity (continued ≥ 1 increase in NCI neurological toxicity grade) and MRI evidence of toxicity (defined as subcortical stripes or increased signal on unenhanced T_1-weighted sequences in peritumoral brain tissue that took up to 1 month after treatment to develop and were suggestive of thrombosis in normal vasculature) were determined to be the dose-limiting toxicities in the phase I study. MRI evidence of toxicity occurred in eight of the first twenty-one patients but was not evaluated in the final seven patients; it was noted only after the first infusion for six of the eight patients. The other two patients had MRI evidence of toxicity noted only after the second infusion. Only one patient had an additional follow-up evaluation to determine the status of the MRI evidence of toxicity. For this patient, the toxicity noted after the first infusion was no longer observed over three subsequent examinations up to 122 days later. Of the eight patients with MRI evidence of toxicity, five had sustained neurologic deficits; the remaining three patients reported no AEs.

Total doses greater than 28 μg, and concentrations greater than 0.7 μg/ml resulted in a notable increase (approximately three-fold) in the number of sustained neurological deficits when compared with patients receiving doses below 28 mg or concentration ≤0.7 mg/ml. The MTD was determined to be 26.8 μg (40 ml of 0.67 μg/ml); 9 of the 12 sustained neurological deficits (in nine patients) occurred at a higher concentration. Three sustained neurological deficits occurred at a lower dose and concentration but none of these were considered related to Tf-CRM107 infusion.

Based on these phase I study results, an infusion of 40 ml of 0.67 μg/ml giving a dose of 26.8 μg per infusion was selected for the phase II study. Because the majority of all laboratory abnormalities began during the infusion and resolved within 2 to 4 weeks after the start of the infusion, it was decided to include two infusions spaced 4 to 10 weeks apart in the phase II study, as well as to include follow-up laboratory measurements at 10 days after initiation of each infusion to closely monitor any laboratory changes. This period also allows time for the patients to recover from biopsy, catheter placement, and the previous infusion.

Phase II clinical study

The phase II clinical study of regional therapy with Tf-CRM107 in patients with refractory and progressive glioblastoma multiforme or anaplastic astrocytoma was completed at nine centers in the U.S. The objectives of the study were to evaluate the efficacy of intratumoral/interstitial regional therapy with Tf-CRM107 in patients with refractory and progressive GBM or anaplastic astrocytoma (AA) and to further evaluate the safety of intratumoral/interstitial regional therapy with Tf-CRM107. Secondary objectives were to evaluate and compare possible differences in efficacy between histological types (GBM or AA), degree of transferrin receptor expression in tumor tissue, and serum anti-DT antibody titer levels and to estimate survival.

This was a multicenter, open-label, single-treatment arm study in which each patient received two separate infusions of Tf-CRM107 between 4 and 10 weeks apart. On each treatment occasion, the total dose of Tf-CRM107 was to be 40 ml of solution at a concentration of 0.67 μg/ml (resulting in a total expected dose of Tf-CRM107 of 26.8 μg during each treatment). The infusion was to be delivered continuously by CED over a period of 5 to 7 days via two catheters stereotactically implanted in the tumor(s). The permitted maximum rate of administration was 0.20 ml/h through each catheter (i.e., total of 0.40 ml/h) up to a total of 40 ml.

Forty-four patients, with a diagnosis of refractory progressive high grade glioma who met all entrance criteria were enrolled. Patients were aged between 18 and 75 years (mean 52 years). All patients had previously undergone conventional treatment (biopsy or debulking surgery and radiation therapy) but their disease was still shown to be progressive.

All patients had a baseline evaluation consisting of blood and urine analyses, neurological testing, Karnofsky Performance Scale Score, and an unenhanced and gadolinium-enhanced MRI scan performed within 24 h to 14 days prior to infusion with Tf-CRM107.

Once admitted to hospital, a CT-guided biopsy of the tumor was performed, the tumor was histologically evaluated, and two silastic catheters were inserted into the tumor. If a satellite tumor was present, one catheter could be placed in the primary tumor and one was placed in the satellite. On the following day, the infusion of Tf-CRM107 (Treatment 1) was initiated through the two catheters at an escalating rate up to 0.20 ml/h for each catheter. Over the next 5 to 7 days, Tf-CRM107 was continuously infused until a total volume of 40 ml was delivered. The catheters were removed after the infusion was completed. Each patient was observed for up to 24 h postinfusion and discharged. At 10 days after initiation of infusion of Tf-CRM107, the patient returned to the hospital for physical and neurological examinations, Karnofsky Performance Scale Score, AE occurrence, and blood/urine tests. The patient also had an additional follow-up visit at 16 to 18 days after the start of the infusion to assess clinical progress and AE occurrence.

Between 4 and 10 weeks after discharge from Treatment 1, the patient was again admitted to the hospital and the same procedure as used in the first treatment regimen was repeated (Treatment 2), including the evaluations at 10 and 16 to 18 days after initiation of the second infusion. Following a protocol amendment, patients who experienced serious cerebral edema during or after their first infusion of Tf-CRM107 were not given a second infusion.

Patients were followed-up at monthly intervals for the first 6 months, and then at 3-monthly intervals for a further 6 months after completion of Treatment 2 (i.e., a total follow-up period of 12 months). Additionally, unenhanced and enhanced MRIs were performed at the 2-, 4-, 6-, 9-, and 12-month follow-up visits.

Unenhanced and gadolinium-enhanced MRIs (TI-weighted) were performed on the same machine for each patient at each specified time point. If considered relevant, T2-weighted images were also obtained. All technical factors, including image orientation, were identical for both the unenhanced and enhanced images. For at least 7 days prior to each MRI, the patient's steroid dose was kept constant. Tumor volume was calculated from contiguous slices (i.e., no interslice gap) according to the central image analysis using software provided by the manufacturer of each MR machine.

Efficacy responses for the tumor(s) treated were categorized according to a scale modified from McDonald et al. (1990) as follows:

Complete response: Complete disappearance of all contrast-enhancing tumor in the area of treatment on MRI; tapering of steroid dose to physiological levels; and normalization (return to baseline) of all signs and symptoms of disease for at least 28 days. (Provided that all other criteria were met, patients on steroids for other indications were evaluated as complete responders.)

Partial response: ≥50% reduction in volume of contrast-enhancing tumor for all treated sites on MRI; stabilized or reduced steroid requirements compared to baseline; and the absence of new or progressing lesions in the area of treatment for a minimum of 28 days.

Stable disease: ≤50% reduction or ≤25% increase in contrast-enhancing tumor volume on MRI; stabilized or increased steroid requirements compared to baseline; and the absence of new lesions in the area of treatment for a minimum of 28 days.

Progressive disease: ≥ 25% increase in contrast-enhanced tumor volume or the presence of any new lesion in the area of treatment.

Nonevaluable (NE):   Patients who have withdrawn from study due to tumor progression and/or for safety reasons.

This survival time for each patient was defined as the number of days from start of first infusion to the date of death or December 31, 1999, if the patient survived beyond that date.

All patients who received infusions with Tf-CRM107 and had a follow-up MRI performed at least 2 months after the final treatment were considered evaluable for response in relation to efficacy. Of the 44 patients 31 (70%) completed both treatment phases (i.e., Treatments 1 and 2). Of the 44 patients 34 (77%) were evaluated for efficacy based on investigator assessments of tumor response; the remaining 10 were considered as not evaluable. A total of five (11.4%) complete responses and seven (15.9%) partial responses were recorded. If all 44 patients are considered the response rates are 11 and 16% for complete and partial responders, respectively.

The median and mean survival times were 37 and 45 weeks, respectively, for all 44 patients. A total of 13 (30%) of the patients survived beyond 12 months from the time of first infusion and one patient was still alive after 5 years.

All of the 44 patients who enrolled in the study received at least one infusion of Tf-CRM107 and were therefore considered evaluable for safety.

In 25 patients 57 serious adverse events (SAEs) were reported during the study; 8 cases of cerebral edema were reported in six patients.

Of the 270 AEs that began during the study, 15 (in 11 patients) were associated with an outcome of death. These three events (20%; all cerebral edema) in two patients were judged to have a likely relationship to Tf-CRM107. One patient had two occurrences of cerebral edema with an onset within 5 days of the Tf-CRM107 infusions. The first infusion was associated with a neurological deficit secondary to cerebral edema, which was judged as likely to be related to Tf-CRM107. The patient underwent a second infusion, also was associated with cerebral edema, which was thought likely to be related to Tf-CRM107. The cause of death in this patient was reported as brain edema, brain tumor, and transtentorial herniation. The investigator did not feel that the cerebral edema was immediately life threatening and considered that it could have been reversed with aggressive treatment. However the patient's family elected to withdraw all support and the patient died two days later. The investigator did not consider that the death was related to Tf-CRM107. The second patient received a single infusion of Tf-CRM107 and had two episodes of cerebral edema, which was judged to have a likely relationship to the study drug. The patient died from respiratory arrest secondary to his brain tumor progression and the death was judged unlikely to be related to Tf-CRM107 by the investigator. Of the 29 poststudy SAEs, 24 events (in 23 patients) were reported to be associated with an outcome of death but none were judged to be related to Tf-CRM107.

Overall, the adverse event profile for Tf-CRM107 was as expected, based on this study population's underlying disease. The most clinically significant adverse events involved cerebral edema, which in a number of cases was serious and required aggressive therapy. The cause of this effect is not clear but may be related to the tumoricidal effect of Tf-CRM107 on the tumor, volume of fluid infused, or an inflammatory response.

Three patients experienced laboratory abnormalities that were reported as AEs but none of these was considered related to Tf-CRM107. There were some transient elevations in systolic blood pressure, which appeared to be related to the procedure.

Conclusions and future directions

The prognosis for relapsing patients with malignant glioma remains poor and there is no standard of care in this situation. Repeat surgery or radiotherapy is often not feasible at relapse and chemotherapy is often given with little expectation of success. A variety of agents have been used, e.g., nitrosoureas, temozolamide, platinum compounds, procarbazine, or combination therapy such as PCV (procarbazine, lomustine, and vincristine). The median survival of recurrent GBM following chemotherapy in small clinical trials is reported, in the literature, to be approximately 26 weeks (Huncharek and Muscat 1998: 1303).

These phase I and II clinical trials with Tf-CRM107 demonstrate that the drug is able to induce responses in a significant number of patients with high-grade gliomas that have failed conventional therapy. The median survival of 37 weeks in the phase II study compares favorably with other therapies that have been used in this patient population (Bower et al. 1997: 484). The majority of adverse events were neurological and reversible. The commonest serious adverse event was cerebral edema. In most patients the edema was controlled by standard management including steroid therapy. The mechanism of cerebral edema is unclear but may be related to the tumoricidal effect of Tf-CRM107 on the tumor, the volume of fluid infused or an inflammatory response or all three of these mechanisms. Prophylactic use of dexamethasone and aggressive management of edema should reduce this complication in future studies.

MRI evidence of toxicity (cortical striping on unenhanced T1 sequences) has been seen in some patients and may represent venous thrombosis. This may be due to the higher expression of TfRs on the surface of capillary endothelial cells and the target specificity of Tf-CRM107. However, despite the presence of cortical striping some patients achieved a complete or partial response with good survival, therefore the clinical relevance of this phenomenon is unclear. Hagihara et al. (2000) reported a study in rats using systemic chloroquine as a potential vascular protective agent. Chloroquine was shown to block the toxicity of DT by increasing and then neutralizing endosomal pH and to block the toxicity of Tf-CRM107, presumably via the same mechanism (Leppla et al. 1980: 2247). In this study the MTD of Tf-CRM107 given by intracerebral infusion was increased from 0.2 to 0.3 μg. These findings raise the possibility that some of the adverse events associated with Tf-CRM107 may be reduced with systemic administration of chloroquine while allowing greater doses of Tf-CRM107 to be delivered to the tumor. Interestingly, chronic administration of chloroquine has recently been shown to enhance the response of GBM tumors to anti-neoplastic therapy possibly by its anti-mutagenic action preventing the appearance of resistant malignant clones (Briceno 2003: 1).

Studies with Tf-CRM107 reported to date have focused on supratentorial tumors. Recently it has been reported that CED can be successfully used to perfuse primate brainstems safely and effectively with macromolecules (Lonser et al. 2002: 905). If this safety is confirmed in humans it may be possible to treat intrinsic tumors of the brainstem including brainstem gliomas, which have been difficult or impossible to treat in the past.

Localized delivery of targeted toxins has a promising future in the treatment of malignant brain tumors. A randomized phase III study of Tf-CRM107 versus best standard care in patients with relapsed and/or progressive unresectable GBM has now commenced at centers in the US, Europe and Israel.

References

American Cancer Society, *Cancer Facts and Figures 2002*.

Angelova-Gateva, P. (1980) "Iron transferrin receptors in rat and human cerebrum," *Agressologie* 21: 27–30.

Bobo, R.H., Laske, D.W., Akbasak, A., Morrison, P.F., Dedrick, R.L., and Oldfield, E.H. (1994) "Convection-enhanced delivery of macromolecules in the brain," *Proceedings of the National Academy of Science (USA)* 91: 2076–2080.

Bower, M., Newlands, E.S., Bleehen, N.M., Brada, M., Begent, R.J., Calvert, H., Colquhoun, I., Lewis, P., and Brampton, M.H. (1997) "Multicentre CRC phase II trial of temozolamide in recurrent or progressive high-grade glioma," *Cancer Chemotherapy Pharmacology* 40: 484–488.

Briceno, E., Reyes, S., and Sotelo, J. (2003) "Therapy of glioblastoma multiforme improved by the antimutagenic chloroquine," *Neurosurgical Focus* 14(2): 1–6.

Connor, J.R. and Fine, R.E. (1986) "The distribution of transferrin immunoreactivity in the rat central nervous system," *Brain Research* 368: 319–328.

Faulk, W.P., His, B.L., and Stevens, P.J. (1980) "Transferrin and transferrin receptors in carcinoma of the breast," *Lancet* 2: 390–392.

Fitzgerald, D.J. and Pastan I. (1989) "Targeted toxin therapy for the treatment of cancer," *Journal of the National Cancer Institute* 81: 1455–1463.

Frankel, A., Kreitman, R.J., and Sausville, E.A. (2000) "Targeted Toxins," *Clinical Cancer Research* 6: 326–334.

Galbraith, G.M.P., Galbraith, R.M., and Faulk, W.P. (1980) "Transferrin binding by human lymphoblastoid cell lines and other transformed cells," *Cellular Immunology* 49: 215–222.

Gatter, K.C., Brown, G., Trowbridge, I.S., Woolston, R.E., and Mason, D.Y. (1983) "Transferrin receptors in human tissues: their distribution and possible clinical relevance," *Journal of Clinical Pathology* 36: 539–545.

Gilliland, D.G., Steplewski, Z., Collier, R.J., Mitchell, K.F., Chang, T.H., and Koprowski, H. (1980) "Antibody-directed cytotoxic agents: Use of monoclonal antibody to direct the action of toxin A chains to colorectal carcinoma cells," *Proceedings of the National Academy of Science (USA)* 77: 4539–4543.

Glioma Meta-analysis Trialists Group (2002) "Chemotherapy in adult high-grade glioma: a systematic review and meta-analysis of individual patient data from 12 randomised trials," *Lancet* 359: 1011–1018.

Greenfield, L., Johnson, V.G., and Youle, R.J. (1987) "Mutations in diphtheria toxin separate binding from entry and amplify immunotoxin selectivity," *Science* 238: 536–539.

Grieg, N.H., Ries, L.G. Yancik, R, and Rapoport S.I. (1990) "Increasing annual incidence of primary malignant brain tumours in the elderly," *Journal of the National Cancer Institute* 82: 1621–1624.

Hagihara, N., Walbridge, S., Olson, A.W., Oldfield, E.H., and Youle, R.J. (2000) "Vascular protection by chloroquine during brain tumour therapy with Tf-CRM107," *Cancer Research* 60: 230–234.

Hamilton, T.A., Weiel, J.E., and Adams, D.O. (1984) "Expression of the transferrin receptor in murine peritoneal macrophages is modulated in the different stages of activation," *Journal of Immunology* 132: 2285–2290.

Hill, J.M., Ruff, M.R., Weber, R.J., and Pert C.B. (1985) "Transferrin receptors in rat brain: Neuropeptide-like pattern and relationship to iron distribution," *Proceedings of the National Academy of Science (USA)* 82: 4553–4557.

Huncharek, M. and Muscat, J. (1998) "Treatment of recurrent high grade astrocytoma; Results of a systematic review of 1,415 patients," *Anticancer Research* 18: 1303–1312.

Jansen, F.K., Blythman, H.E., Carriére, D., et al. (1982) "Immunotoxins: hybrid molecules combining high specificity and potent cytotoxicity," *Immunological Reviews* 62: 185–216.

Jefferies, W.A., Brandon, M.R., Hunt, S.V., Williams, A.F., Gatter, K.C., and Mason, D.Y. (1984) "Transferrin receptor on endothelium of brain capillaries," *Nature* 312: 162–163.

Johnson, V.G., Wilson, D., Greenfield, L., and Youle, R.J. (1988) "The role of the diphtheria toxin receptor in cytosol translocation," *Journal of Biological Chemistry* 263: 1295–1300.

Johnson, V.G., Wrobel, C., Wilson, D., et al. (1989) "Improved tumour-specific immunotoxins in the treatment of CNS and leptomeningeal neoplasia," *Journal of Neurosurgery* 70: 240–248.

Klausner, R.D., VanRenswoude, J., Ashwell, G., et al. (1983) "Receptor-mediated endocytosis of transferrin in K562 cells," *Journal of Biological Chemistry* 258: 4715–4724.

Larrick, J.W. and Cresswell, P. (1979) "Modulation of cell surface iron transferrin receptors by cellular density and state of activation," *Journal of Supramolecular Structure* 11: 579–586.

Lashford, L.S., Davies, A.G., Richardson R.B., et al. (1988) "A pilot study of ^{131}I mono-clonal antibodies in the therapy of leptomeningeal tumours," *Cancer* 61: 857–868.

Laske, D.W., Ilercil, O., Akbasak, A., Youle, R.J., and Oldfield, E.H. (1994) "Efficacy of direct intratumoural therapy with targeted protein toxins for solid human gliomas in nude mice," *Journal of Neurosurgery* 80: 520–526.

Leppla, S., Dorland, R.B., and Middlebrook, J.L. (1980) "Inhibition of diphtheria toxin degradation and cytotoxic action by chloroquine," *Journal of Biological Chemistry* 255: 2247–2250.

Lonser, R.R., Walbridge, S., Garmestani, K., Butman, J.A., Walters, H.A., Vortmeyer, A.O., Morrison, P.F., Brechbiel, M.W., and Oldfield, E.H. (2002) "Successful and safe perfusion of the primate brainstem: *in vivo* magnetic resonance imaging of macromolecular distribution during infusion," *Journal of Neurosurgery* 97: 905–913.

McDonald, D.R., Cascino, T.L., Schold, S.C. Jr, and Cairncross, G. (1990) "Response criteria for phase II studies of supratentorial malignant glioma," *Journal of Clinical Oncology* 8: 1277–1280.

Oldfield, E.H., Laske, D.W., and Youle RJ (1992) Physicians' IND number BB4570.

Salcman, M. (1990) "Epidemiology and factors affecting survival," in M.L. Apuzzo (ed.), *Malignant Cerebral Glioma*, Park Ridge III: American Association of Neurological Surgeons, 95–109.

Schold, S.C. Jr., Burger, P.C., Mendolsohn D.B., et al. (1997) *Primary Tumours of the Brain and Spinal Cord*, Boston MA, Butterworth-Heinemann, 51–54.

Shindelman, J.E., Ortmeyer, A.E., and Sussman, H.H. (1981) "Demonstration of the transferrin receptor in human breast cancer tissue: Potential marker for identifying dividing cells," *International Journal of Cancer* 27: 329–334.

Trowbridge, I.S. and Domingo, D.L. (1981) "Anti-transferrin receptor monoclonal antibody and toxin-antibody conjugates affect growth of human tumour cells," *Nature* 294: 171–173.

Trowbridge, I.S., Newman, R.A., Domingo, D.L., and Sauvage, C. (1984) "Transferrin receptors: structure and function," *Biochemical Pharmacology* 33: 925–932.

Uhr, J.W. (1984) "Immunotoxins: harnessing nature's poisons," *Journal of Immunology* 133: i–x.

Youle, R.J. and Colombatti, M. (1986) "Immunotoxins: monoclonal antibodies linked to toxic proteins for bone marrow transplantation and cancer therapy," in J.A. Roth (ed.), *Monoclonal Antibodies in Cancer; Advances in Diagnosis and Treatment*, New York, Futura Publishing Co., 173–213.

Zovickian, J., Johnson, V.G., and Youle, R.J. (1987) "Potent and specific killing of human malignant brain tumour cells by an anti-transferrin receptor antibody-ricin immunotoxin," *Journal of Neurosurgery* 66: 850–861.

chapter eight

GM-CSF receptor-targeted therapy of human leukemia

Arthur E. Frankel, Philip D. Hall, Dongsun Cao, Tie Fu Liu, Marlena Moors, Kimberley A. Cohen, Andrew M. Thorburn, and Robert J. Kreitman

Contents

Introduction

Patients with chemotherapy refractory acute myeloid leukemia (AML) respond poorly to further cytotoxic drugs and have a median survival of less than 6 months. The leukemic stem cells from these patients often show multidrug resistance to a variety of agents that modify DNA synthesis or cell proliferation. Novel agents that kill leukemic cells by different mechanisms are needed. One such new class of antileukemia drugs are diphtheria fusion proteins, which consist of the catalytic and translocation domains of diphtheria toxin (DT) fused to leukemic blast selective ligands. Our laboratory has focused on the synthesis and testing of diphtheria fusion proteins for

AML. In this chapter, we review our preclinical and clinical experiences with $DT_{388}GMCSF$ prepared by genetically linking DT_{388} to human granulocyte-macrophage colony stimulating factor (GMCSF).

AML

AML is the most common acute leukemia in adults and the second most common leukemia in children.[1] There are 10,000 estimated new cases per year in the U.S. With the prolonged hospitalizations associated with treatment and complications, the disease represents a significant share of health care costs. Two types of drugs have shown significant activity in AML including cytosine arabinoside and the topoisomerase inhibitors — anthracyclines, amsacrine, and etoposide. With combination induction and consolidation chemotherapy, complete remission rates of about 70% have been achieved.[2] However, most patients ultimately relapse and die from the disease or complications of treatment.

The prognosis is dismal for patients with relapsed or refractory AML. Except for the minority of patients who undergo allogeneic bone marrow transplants or have initial complete remissions of >12 months, patients receive similar salvage therapy to that used in their induction/consolidation and have median survival of weeks to months.[3] Two-year survival in this subgroup of relapsed and refractory patients is rare (Figure 8.1).[4] The mainstay of therapy for relapsed AML other than allogeneic stem cell transplant is

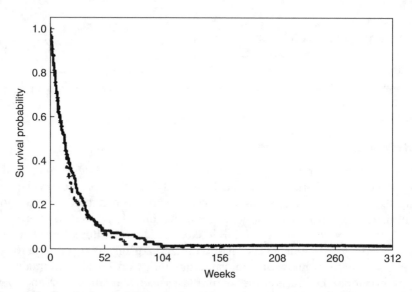

Figure 8.1 Survival probability of relapsed or refractory AML patients whose initial complete remission was <1 year. ---, investigational agents (171 patients); —, high-dose cytarabine (176 patients).[4] These patients represent 72% of the total relapsed/refractory patients

high-dose cytarabine. Other salvage regimens that have been tested include anti-CD33 antibody–calicheamicin conjugate (Mylotarg®), aclarubicin + etoposide, subcutaneous cytarabine + etoposide, cyclophosphamide + etoposide, interleukin-2, topotecan, thalidomide, gemcitabine, troxacitabine, anti-CD33 antibody-[213]Bi conjugate, temozolomide, and Zarnestra farnesyl transferase inhibitor. None have produced durable remissions in the majority of relapsed AML patients.[5–16]

Chemotherapy-resistant blasts contribute to treatment failure in AML patients.[17] These blasts are resistant to numerous cytotoxic agents including those to which the patient has not been previously exposed. This multidrug resistance has been linked to altered expression of one or more resistance proteins, which may influence drug efflux (P-glycoprotein and lung resistance protein), drug metabolism (glutathione S-transferase and metallothionein), substrate levels (thymidylate synthetase and topoisomerase II), or cell death regulation (Bcl2 and p53). Prospective and retrospective clinical studies have shown a worse prognosis for de novo AML with resistance phenotypes due to abnormal concentrations of these molecules.[18–24] Further, patients with abnormal levels of more than two of these factors had significantly higher relapse rates.[25] Since most of the multidrug resistance phenotypes target small molecular weight inhibitors of DNA synthesis or cell proliferation, investigators have sought agents that induce leukemia cell death by other than damage to DNA or cell division. An added benefit of such new agents would be the possibility of distinct, nonoverlapping toxicities with current cytotoxic chemotherapy permitting combinations.

AML-targeted toxins

Targeted toxins consist of protein synthesis inactivating peptide toxins covalently linked to tumor-selective ligands. Several toxins targeted to AML have been made with monoclonal antibodies reactive to the AML cell-surface antigens — CD13, CD14, CD33, and CD71.[26–29] In addition, fusion proteins using the AML blast ligand granulocyte-macrophage colony-stimulating factor were made.[30,31] The toxins used included the plant toxins (ricin, gelonin, saporin) and the bacterial toxin (*Pseudomonas* exotoxin). In each case, the chimeric toxins had low potency on myeloid cells. These toxins need to route after internalization to the endoplasmic reticulum in order for translocation to the cytosol.[32] However, in myeloid cells, most endocytosed proteins are rapidly trafficked to lysosomes where they are degraded. A different toxin is needed, which can escape to the cytosol from AML blast early endosomes.

DT fusion proteins

DT enters the cytosol from an early endosomal compartment. Thus, it may avoid premature lysosomal degradation by myeloid cells. DT is a 535-amino acid protein with three domains consisting of a catalytic domain also called the A fragment (amino acids 1–186) connected by a 14-amino acid arginine-rich disulfide loop to a translocation domain containing multiple amphipathic helices (amino acids 187–381) followed by a flexible linker peptide (amino acids 382–390) and a β-sheet-rich cell-binding domain (amino acids 390–535).[33]

DT binds via amino acid residues in the cell-binding domain including Lys-516 to a crevice (amino acid residues 122–148 including Glu-141) in the extracellular epidermal (EGF)-like domain of cell-surface expressed heparin-binding epidermal growth factor-like (HB-EGF) precursor in association with CD9 and heparin sulfate proteoglycan.[34] The complex undergoes clathrin and dynamin-dependent endocytosis.[35] In early endosomes, DT undergoes low pH-induced protonation of the translocation domain helical hairpin carboxylates (Glu-349 and Asp-352 of the TH8-TH9 hairpin), insertion of the TH8 and TH9 amphipathic helices into the vesicle membrane, furin cleavage at the arginine-rich loop, unfolding of the catalytic domain, reduction of the disulfide bridge linking the A fragment to the remainder of DT, and transfer of the A fragment to the cytosol.[36] In the cytosol, the A fragment including residues 39–46 and Glu-148 catalyze the ADP-ribosylation of the diphthamide residue in domain IV of elongation factor 2 (EF2).[37] The modified EF2 cannot displace the tRNA–peptidyl complex from the A site to the P site of the ribosome, and cellular protein synthesis is halted. Cells undergo lysis or programmed cell death.[38]

New targeting specificities for DT have been achieved by genetically replacing the C-terminal receptor-binding domain (amino acid residues 391–535) with cell selective ligands such as melanocyte-stimulating hormone, interleukin-2, interleukin-6, gastrin-releasing peptide, and others.[39] These DT fusion proteins have shown potent and selective killing to receptor positive cells in tissue culture, animal models, and patients.

GMCSF receptor

GMCSF is a multifunctional 124-amino acid cytokine important in the proliferation and differentiation of myeloid progenitors.[40] The crystal structure of GMCSF shows that it folds into an antiparallel four-helix bundle.[41] The receptor for GMCSF has been identified on myeloid progenitors, mature monocytes, granulocytes, macrophages, and on myeloid leukemias.[42,43] GMCSF residues at both the N-terminus and C-terminus bind the receptor. While the GMCSF receptor on leukemic blasts undergoes rapid receptor-mediated endocytosis after ligand binding, this does not occur as rapidly with ligand binding to normal progenitor receptors.[43] Based on the limited normal tissue distribution of its receptor and its efficient internalization by myeloid blasts, we chose GMCSF for fusion to the enzymatic and translocation domains of DT.

$DT_{388}GMCSF$

$DT_{388}GMCSF$ was produced by genetic engineering. Briefly, the pET3a plasmid, which has the T7 promoter, pMB1 replicator, and β-lactamase gene, was digested with restriction enzymes and DNA was inserted encoding methionine, amino acids 1–388 of DT, a histidine–methionine linker, and amino acids 1–124 of human GMCSF.[31] This plasmid, pRKDTGM, and pUBS500 (which has a kanamycin resistance gene and encodes the *dna Y* gene product — tRNA$^{Arg}_{AGA/AGG}$) were used to transform BL21(DE3) competent *Escherichia coli.*[44] Recombinant protein was induced with isopropyl-β-D-thiogalactopyranoside (ITPG). Cells were extracted and inclusion bodies were washed, denatured with guanidine hydrochloride and dithioerythritol. Protein was refolded at 4 °C in an arginine/Tris/EDTA/oxidized glutathione buffer. After dialysis, DT$_{388}$GMCSF was purified by anion exchange, size exclusion, and polymixin B affinity chromatography. Protein was aseptically vialed at 1.5 mg/ml in phosphate-buffered saline and stored at −80°C until used.

DT$_{388}$GMCSF was characterized chemically. A model of the molecule is shown in Figure 8.2. The protein was 99% pure by gel electrophoresis, and there were <1% aggregates by high pressure liquid chromatography. There were low levels of endotoxin (1.7 eu/mg) and bacterial DNA contamination (<113 pg/mg). The DT$_{388}$GMCSF affinity for the GMCSF receptor was 1.3×10^9 M^{-1}/l. This compares favorably to the GMCSF affinity for the GMCSF receptor, which was 6.7×10^9 M^{-1}/l (five-fold higher). The ADP-ribosylating activity was identical to CRM107 DT. The molecular weight determined by tandem mass spectroscopy was 57,232 Da, which was similar to the calculated molecular weight of 57,082 Da. There were no free thiols (≤0.17/molecule), and the protein had a pI of 6.0 by isoelectric focusing. The molecule reacted with antibodies to DT and GMCSF by immunoblots.

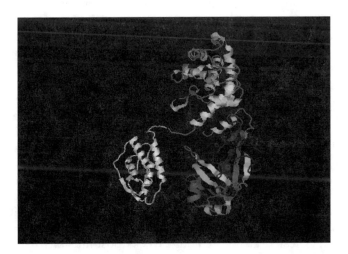

Figure 8.2 Ribbon diagram of a three-dimensional structure for the α-carbon backbone of DT$_{388}$GMCSF. The lower right shows the DT catalytic domain; the upper right shows the DT translocation domain; the left shows the GMCSF

Preclinical studies with DT$_{388}$GMCSF

DT$_{388}$GMCSF prepared in our laboratory and similar compounds made by others were toxic to AML cell lines and patient leukemic progenitors.[45–50] Further, cytotoxicity to chemotherapy-resistant cell lines and therapy-refractory AML patient progenitors was also observed.[51] Significantly, DT$_{388}$GMCSF was not toxic to normal human myeloid progenitors.[31,47,52,53] When DT$_{388}$GMCSF was combined with cytotoxic drugs, synergistic cell killing of AML cell lines was seen.[54,55] We recently examined factors predicting DT$_{388}$GMCSF sensitivity of patient leukemic progenitors and found that neither receptor density nor receptor internalization correlated with cell-kill (Figure 8.2). Cells intoxicated by the fusion protein showed programmed cell death, which was FADD but not death receptor dependent suggesting a novel apoptotic pathway.[56] Because of liver toxicity observed in the clinical trial described later, assays of fusion protein cytotoxicity to hepatocyte cell lines and primary cultures were undertaken. No damage to human or rat hepatocytes was observed after incubation with DT$_{388}$GMCSF or DT$_{388}$mGMCSF, respectively.

Animal studies with DT$_{388}$GMCSF have been done in mice, rats, and cynomolgus monkeys. Mice treated with the human GMCSF receptor-targeted fusion protein by intraperitoneal injections daily for 5 days tolerated a daily dose of 84 μg/kg/day.[57] The dose-limiting toxicity was acute proximal tubular necrosis. The drug had a peak level at 40 min after intraperitoneal injection and a circulating half-life of 24 min. The circulating drug concentrations exceeded levels needed to kill leukemic blasts for over 5 h. Only 24% of mice developed a weak immune response 3 weeks after treatment. Severe combined immunodeficient (SCID) mice inoculated with HL-60 human AML cells develop leukemia with abdominal masses, infiltration of the liver and bone marrow, and peripheral blasts and have a median survival of 42.5 days.[58] Treatment of these animals with five daily intraperitoneal injections of 84 μg/kg/day DT$_{388}$GMCSF significantly prolonged survival to a median of 83 days ($p < 0.001$), and six of fifteen treated animals had no visible disease at >150 days. No evidence of microscopic disease based on histopathology was observed in three of these animals. Similar efficacy in mouse leukemia models was reported by others.[59,60] Cynomolgus monkeys have GMCSF receptors that cross-react with DT$_{388}$GMCSF. Monkeys treated with five daily intravenous bolus infusion of 10 μg/kg/day DT$_{388}$GMCSF developed dose-limiting neutropenia and sepsis.[61] Necropsies of these animals showed no damage to organs other than the bone marrow including kidney, heart, liver, central nervous system, or lung. The maximal tolerated dose (MTD) was 7.5 μg/kg/day DT$_{388}$GMCSF for 5 days where only transient myelosuppression and hypoalbuminemia were seen. The fusion protein half-life in the circulation was 30 min. Immune responses were minimal in all monkeys tested at both 2 and 4 weeks postinfusions with antibody concentrations of <1 μg/ml. Similar results were reported by others.[59]

To better explore receptor-specific toxicities of DT$_{388}$GMCSF in a model more amenable to study than monkeys, we evaluated DT$_{388}$mGMCSF in rats.

At intraperitoneal doses of 37.5 µg/kg/day for 3 to 7 days, transaminasemia, disappearance of liver Kupffer cells, and appearance of apoptotic cells in liver sinusoids were seen. Significant mortality was observed at this and higher doses, but there was no evidence of severe liver injury nor sepsis on necropsies. Liver enzyme elevations after $DT_{388}mGMCSF$ administration to rats has been previously reported.[62] Coadministration of steroids with $DT_{388}mGMCSF$ worsened the rat transaminasemia in our study. Circulating rat interleukin-18 levels increased, but administration of glycine or an interleukin-18 inhibitor (2357) failed to prevent the rat transaminasemia.

Clinical studies with $DT_{388}GMCSF$

Thirty-seven patients with AML have been treated with $DT_{388}GMCSF$ for a total of 46 courses (nine patients received a second course).[63] The mean age was 54 years (range, 12 to 84), and there were 16 males and 21 females (Table 8.1). Although the eligibility criteria stipulated age >18 years, one child was allowed under a Federal Drug Administration–approved exemption. Four patients were in first relapse; two patients were in second relapse; and thirty-one patients had refractory AML. Seven patients had undergone autologous bone marrow transplants; six had previously received allogeneic bone marrow transplants, and twenty-four had not received transplants. The cytogenetics were poor risk in seventeen patients, intermediate risk in eleven patients, and good risk in two patients. In seven patients, cytogenetics were not performed. Prior myelodysplasia had been present in seven of the thirty-seven patients.

Patients were treated at one of six dose levels of $DT_{388}GMCSF$ (1, 2, 3, 4, 4.5, or 5 µg/kg/day) for up to 5 days. The first nine patients were treated at 1 or 2 µg/kg/day with premedications consisting of acetaminophen and diphenhydramine. Because of infusion-associated fevers and chills, the next twenty-four patients were treated at 3, 4, 4.5, or 5 µg/kg/day $DT_{388}GMCSF$ with the addition of corticosteroids to the premedication (solumedrol intravenously 0.5 mg/kg every 6 h followed by a taper). Based on the hypothesis that steroids worsened liver injury from the DT fusion protein, the last four patients were treated at 3 or 4 µg/kg/day without corticosteroid premedication. The fusion protein was administered over 15 min by slow bolus intravenous infusion. Patients also received at least 1 l of intravenous fluids daily. After one month or longer, nine patients received a second course.

The dose-limiting toxicity for $DT_{388}GMCSF$ plus steroids was liver injury. Nine patients were treated at the 4.5 to 5 µg/kg/day dose levels, and two of these patients had severe hepatic toxicity. Patient 28 developed liver failure and died 1 week after treatment. This was coincident with florid leukemic progression. Necropsy showed centrilobular hepatic necrosis. Patient 31 developed transient hepatic encephalopathy with elevated ammonia levels 4 days after therapy. The liver dysfunction resolved, but the patient developed renal failure contributed to by aminoglycosides, amphotericin, and high salt formulated $DT_{388}GMCSF$. Without dialysis, the patient died 1 week later.

Table 8.1 Clinical characteristics of DT$_{388}$GMCSF-treated AML patients

Patient no.	Age year/ Sex	Disease status (year from diagnosis)	Treatment history	Cytogenetics	Circulating blast count (/μl)	Bone marrow blast %
1	69/F	Ref. (2)	Hydrea;Topo	46XiX(p10)	0	5
2	55/M	1st Rel. (1)	AlloBMT;DLI;Ara-C	Del7(q22)	1,127	6
3	66/F	1st Rel. (1)	Ida/Ara-Cx2	+8,+11,t(1;16)	0	10
4	46/F	2nd Rel. (5)	Dauno/Ara-C;Ara-C;AutoBMT	Normal	72	32
5	69/F	Ref. (1)	Mito/Ara-C;Ara-C;CTX/VP16	Normal	8,024	10
6	41/F	Ref. (1)	Dauno/Ara-C;Ara-C/VP16;AutoBMT	+21,+22	10,004	98
7	84/F	Ref. (1)	Cytarabine;6TG;Mito;Vinc	ND	684	93
8	61/M	1st Rel. (1)	Ida/Ara-C;Ara-C	ND	96	30
9	76/M	Ref. (0.5)	Ida/Ara-C;Ara-C;CTX/VP16	ND	240	10,LC
10	53/F	Ref. (1)	Ida/Ara-C;Ara-C;Topo	+19,+1-2r	276	30
11	60/M	Ref. (2)	Dauno/Ara-C;Ara-C;Topo;Mito	+13	4,485	87
12	27/F	Ref. (1)	Dauno/Ara-C;Ara-C;VP16/Mito; Topo/Ara-C;Flud/Ara-C	t(1;9)(q21;q21)	15,696	90
13	77/M	Ref. (1)	Dauno/Ara-C;Ara-C;VP16/Mito;Topo/Ara-C; Ida/Ara-C;Ida	Normal	4,324	89
14	24/F	1st Rel. (0.5)	Ida/Ara-C;Ida	+8,+11,+13	0	35
15	45/M	2nd Rel. (2.5)	VP16/Ara-C;Dauno/Ara-C;AutoBMT	ND	616	80
16	25/M	Ref. (1.5)	Dauno/VP16/Ara-C;Ara-C;AutoBMT; Mito/VP16/Ara-C;Topo/Ara-C	Del11(q22), Del5(q12)	0	40
17	52/F	Ref. (1)	Ida/Ara-C;Ara-C;Mito/VP16/Ara-C; VP16/Ara-C;Topo/Ara-C	Normal	6,300	95
18	39/F	Ref. (2)	Dauno/Ara-C;Ara-C;MTX/Ara-C/Ida;CTX	t(8;21)	0	52
19	52/F	3rd Rel. (2)	Mito/Ara-C;Ara-C;AlloBMTx2; VP16/Ara-C;IT Ara-C;WBI	Normal	0	75,LC
20	54/M	Ref. (2)	Dauno/VP16/Ara-C;VP16/Ara-C;Ida/ Ara-C;AutoBMT	-2,-5,+8,+11,+ 16,+21	14,168	60
21	73/F	1st Rel. (1)	Ida/Ara-C;Ara-C	t(1;11),t(2;11),+4,+6	208	88

#	Age/Sex	Dose	Cytogenetics	Prior therapy		Blast %
22	72/F	Ref. (4)	Normal	Ida/Ara-C;Topo/Ara-C	0	28
23	46/M	Ref. (0.5)	+11,Del7	Ida/VP16/Ara-C;Topo/Ara-C	1,588	82
24	32/F	Ref. (1)	ND	Ida/Ara-C;AlloBMT;VP16;CTX	424	14
25	77/M	Ref. (1)	Del7(q22)	Ida/Ara-Cx2	50	15
26	67/M	Ref. (0.5)	ND	Topo;Topo/Ara-C;Mito/Ara-C	0	91
27	61/M	Ref. (1)	Normal	Dauno/Ara-C;VP16/Ara-C;CTX/VP16;AutoBMT	0	23
28	57/F	Ref. (0.5)	t(2;8),Del8,−5,Del7	CTX/Ara-C/Topo	0	11
29	12/F	Ref. (0.5)	+19	Dauno/Ara-C/6TG;Ara-C;AlloBMT	0	87
30	31/M	Ref. (1)	Normal	Dauno/Ara-C;VP16/Ara-C;Flud/CPT11;Gem/CPT11;Mylotarg;CTX/VP16	500	92
31	80/F	Ref. (1)	ND	Dauno/Ara-C/VP16;Ara-C;Mylotarg	60	47
32	19/M	Ref. (1)	t(8;21)	Ida/Ara-C;Ara-C;Flud/Ara-C	400	96
33	60/M	Ref. (0.5)	−7,Del5	Dauno/Ara-C;Ara-C;AutoBMT	0	9
34	34/F	Ref. (2)	t(4;15),Del2,−13,Del5	Ida/Ara-C;AlloBMT;Flud/Ara-C;DLI;Adria;Mylotarg;Flud/Topo/Ara-C	600	22
35	57/F	Ref. (1)	Normal	CTX;Flud;AlloBMT	500	85
36	76/F	Ref. (1)	Normal	Dauno/Ara-C;Mylotarg;Ara-C	4,200	90
37	30/M	Ref. (1)	Normal	Dauno/Ara-C;Ara-C;Mito/VP16;Mylotarg;Flud/Ara-C	4,000	100

LC, leukemia cutis; Ara-C, cytarabine; CTX, cyclophosphamide; VP16, etoposide; Ida, idarubicin; Mito, mitoxantrone; Dauno, daunorubicin; DLI, donor lymphocyte infusion; BMT, bone marrow transplant; TG, thioguanine; Auto, autologous; Allo, allogeneic; r, ring; Flud, fludarabine; IT, intrathecal; WBI, whole brain irradiation; Topo, topotecan; ND, not determined; Ref., refractory; Rel., relapsed; vinc, vincristine; Mylotarg, gemtuzumab ozogamicin; Gem, gemcitabine.

Patient #29 was treated under an IND exemption because of age.

Transient elevations in AST were seen in most patients (Table 8.2). Furthermore, the peak AST level correlated with the dose ($p = 0.002$) and peak $DT_{388}GMCSF$ level ($p = 0.0002$). The AST rose disproportionately relative to ALT and GGT. Steroids, calcium channel blocker, rofecoxib, and antibody to TNF-α did not influence this toxicity. In contrast, patients not receiving steroid prophylaxis had much less evidence of liver injury. No liver biopsies have been performed in $DT_{388}GMCSF$-treated patients, but elevated circulating levels of two inflammatory cytokines, IL-8 and IL-18, were observed. There was a significant positive association of IL-18 levels with AST levels ($p = 0.05$). The elevation of IL-18 occurred earlier (by day 2) than the elevation of IL-8 (day 4 to 8). A transient asymptomatic sinus bradycardia occurred in some patients with steroid prophylaxis that was responsive to oral theophyllines. Echocardiograms, during and after treatment, did not show reduced cardiac function. There were no cardiac enzyme elevations (CPK-MB fractions) and no evidence of heart block by electrocardiography. Asymptomatic, transient hypocalcemia occurred between days 4 and 10. The incidence of hypocalcemia was dose dependent but only occurred with steroid prophylaxis. Neither neurological nor muscular abnormalities were observed. When measured the vitamin D levels were normal and the parathyroid hormone levels were slightly increased. Serum calcium values of <7 mg/dl were treated with intravenous calcium gluconate. Serum calcium values were not corrected for the serum albumin levels. Transient CPK and lactate dehydrogenase elevations were observed in most patients and paralleled the changes in AST. However, there were not signs of muscle injury or RBC injury measured by serum aldolase and haptoglobin or peripheral smear. Vascular leak syndrome (VLS) with the combination of edema, weight gain, hypoxia, hypotension, and hypoalbuminemia was noted in patient #9 and #11. Patient #9 had not been receiving thyroid medications for 1 week, and the symptoms resolved with diuresis and reinstitution of thyroxine. Patient #11 had fluid overload combined with corticosteroids that also resolved with diuresis. The other patients had signs of the components of VLS with weight gain and/or hypoalbuminemia, but there was no associated edema, hypoxemia, or hypotension. The incidence and frequency of toxicities to $DT_{388}GMCSF$ are shown in Tables 8.2 and 8.3. On the basis of the occurrence of grade 4 and 5 liver toxicities in patients #28 and #31, we determined the MTD for the $DT_{388}GMCSF$ plus steroid combination to be 4 μg/kg/day. To assess the MTD of $DT_{388}GMCSF$ alone, four patients have been treated without steroids at the 3 to 4 μg/kg/day dose level. These patients elected to take supplemental glycine 5 g po q4 h to reduce inflammatory responses. Neither DLTs nor liver injury occurred (see Table 8.2, patients #34b to 37) (Figure 8.3). However, transient (< 24 h) fever, chills, nausea, and vomiting occurred.

Pharmacokinetic data were obtained for the first infusion of each course of all patients and for the first and last infusion on five patients. The peak $DT_{388}GMCSF$ serum level occurred at 2 min postinfusion, and the concentration decreased over time exponentially with a $t_{1/2}$ of approximately 30 min. The peak fusion protein concentration was not significantly correlated with dose

Table 8.2 Dose level and drug-related toxic effects of $DT_{388}GMCSF$ in AML patients

Patient no.	Dose (µg/kg/day)	Drug-related side effects (CTC toxicity grade)	Cause of death (day post-therapy)
1	1	Gr[b] 1 myalgias	Progressive disease (21)
2	1	Gr 2 AST;Gr 1 Alk Phos	Progressive disease (213)
3	1	Gr 1 myalgias;Gr 2 AST;Gr 1 Alk Phos	Progressive disease (230)
4	2	Gr 2 fever;Gr 2 hypocalcemia;Gr 2 hypoalbuminemia;Gr 3 AST	Progressive disease (357)
5	2	Gr 1 Alk Phos	Progressive disease (135)
6	2	Gr 1 Alk Phos	Progressive disease (130)
7	2	Gr 2 fever;Gr 2 hypocalcemia;Gr 3 AST	Progressive disease (60)
8	2	Gr 1 fever;Gr 1 AST; Gr 2 hypocalcemia	Progressive disease (94)
9	2	Gr 1 elevated Cr;Gr 1 Bili;Gr 1 AST;Gr 1 Alk Phos;Gr 3 VLS	Progressive disease (45)
10	3	Gr 1 fever;Gr 2 AST	Progressive disease (130)
11	3	Gr 1 weight gain;Gr 2 AST; Gr 2 VLS;Gr 2 hypoalbuminemia; Gr 2 hypocalcemia;Gr 2 Bili;Gr 2 hypotension;Gr 1 Alk Phos	Progressive disease (105)
12	3	Gr 2 AST;Gr 2 hypoalbuminemia;Gr 1 bradycardia	Post-BMT GVHD (72)
13	4	Gr 1 edema;Gr 1 hypoalbuminemia;Gr 2 hypotension;Gr 2 hypocalcemia;Gr 2 AST	Sepsis (13)
14	4	Gr 1 Alk Phos;Gr 2 AST	Post-BMT GVHD (90)
15	4	Gr 1 fever;Gr 2 hypoalbuminemia;Gr 2 GGT;Gr 2 hypocalcemia; Gr 3 CPK;Gr 3 ALT;Gr 4 AST	Alive (700+)
16	4	Gr 1 AST;Gr 2 hypocalcemia;Gr 2 hypoalbuminemia	Progressive disease (90)
17	4	Gr 1 bradycardia;Gr 1 AST;Gr 1 Alk Phos;Gr 1 hypoalbuminemia;Gr 2 hypocalcemia	Progressive disease (90)
18	4	Gr 1 AST;Gr 1 bradycardia	Progressive disease (142)
19c	4	Inevaluable for toxicities	Heart and renal failure (10)
20	4	Gr 1 AST	Progressive disease (34)
21	5	Gr 1 Alk Phos;Gr 1 fever;Gr 2 AST;Gr 2 CPK;Gr 3 hypocalcemia	Progressive disease (37)
22	5	Gr 1 GGT;Gr 3 ALT;Gr 3 CPK;Gr 3 hypocalcemia;Gr 4 AST	Pneumonia (380)
23	5	Gr 1 ALT;Gr 1 CPK;Gr 2 AST;Gr 2 hypocalcemia	CNS hemorrhage (17)

Table 8.2 Continued

Patient no.	Dose (μg/kg/day)	Drug-related side effects (CTC toxicity grade)	Cause of death (day post-therapy)
24	5	Gr 1 Alk Phos;Gr 1 CPK;Gr 2 AST;Gr 2 Bili;Gr 3 GGT;Gr 3 ALT;Gr 3 hypocalcemia	Progressive disease (350)
25	3	Gr 1 Alk Phos;Gr 2 Bili;Gr 2 AST;Gr 2 Cr;Gr 2 CPK;Gr 3 ALT;Gr 3 hypocalcemia	Progressive disease (40)
26	4	Gr 1 bradycardia;Gr 1 CPK;Gr 1 Alk Phos;Gr 1 GGT;Gr 2 hypocalcemia;Gr 2 AST	CNS hemorrhage (9)
27	4.5	Gr 1 Alk Phos;Gr 1 hypoalbuminemia;Gr 2 GGT;Gr 2 AST;Gr 2 CPK;Gr 3 ALT;Gr 3 hypocalcemia	Mylotarg liver failure (60)
28	4.5	Gr 4 ALT;Gr 4 AST;Gr 4 hypocalcemia;Gr 5 hepatic failure;Gr 2 Cr;Gr 4 GGT;Gr 3 CPK;Gr 3 PTT;Gr 2 PT	Liver failure (19)
29	4.5	Gr 1 fibrinogen; Gr 1 PT;Gr 2 Bili;Gr 2 hypoalbuminemia;Gr 3 AST;Gr 3 ALT	Progressive disease (38)
30	4.5	Gr 1 Bili;Gr 1 PT;Gr 1 hyperkalemia;Gr 2 hypocalcemia;Gr 2 fibrinogen	Progressive disease (31)
31	4.5	Gr 1 PT;Gr 2 hypoalbuminemia;Gr 4 renal failure;Gr 4 hepatic encephalopathy; Gr 4 AST;Gr 3 hypocalcemia	Renal failure (18)
32	4	Gr 2 AST;Gr 2 ALT	Post-BMT GVHD (90)
33	4	Gr 2 fever;Gr 3 Cr;Gr 4 hypocalcemia;Gr 2 hypoalbuminemia;Gr 5 renal failure; Gr 4 acidosis;Gr 2 AST	Renal failure (12)
34	3	Gr 3 AST;Gr 2 ALT;Gr 2 Alk Phos;Gr 2 hypoalbuminemia;Gr 3 hypocalcemia;Gr 2 PTT;Gr 1 PT; Gr 2 vomiting	Progressive disease (95)
35	3	Gr 2 AST;Gr 3 fever;Gr 1 PT;Gr 1 PTT;Gr 2 hypocalcemia;Gr 1 nausea;Gr 2 Albumin	Progressive disease (120)
36	4	Gr 1 PTT; Gr 2 hypocalcemia;Gr 2 Albumin;Gr 1 Alk Phos;Gr 2 nausea; Gr 2 vomiting;Gr 2 fatigue;Gr 1 fever	Alive (60+)
37	4	Gr 1 PTT;Gr 1 hypocalcemia;Gr 3 fever;Gr 2 nausea;Gr 2 vomiting	Alive (30+)

Patient # 2, 3, 4, 8, 11, 12, 22, 24, and 34 had two courses, and all toxicities from all courses are listed with highest grade observed.

Gr,grade; Bili, bilirubin; Alk phos, alkaline phosphatase; BMT, bone marrow transplant; GVHD, graft-versus-host disease; CNS, central nervous system; PT, protime; PTT, partial thromboplastin time; Cr, creatinine; GGT, gamma glutamylaminotransferase; CTC, common toxicity criteria.

Patient #19 was inevaluable for toxicities because of heart and kidney dysfunction, which made the patient ineligible for study.

Patients #31 and 33 had renal failure associated with high salt formulation of study drug.

Table 8.3 Relationship of grade 3–5 DT$_{388}$GMCSF + steroid-related toxicities and dose

Dose (μg/kg/day)	No. of patients at each dose	Grade 3–5 drug combination-related toxicities n (%)						
		Transminasemia	VLS	Elevated CPK	Hypocalcemia	Elevated PTT	Renal failure	Hepatic failure
1	3	0	0	0	0	0	0	0
2	6	2(33)	1(16)	0	0	0	0	0
3	5	1(20)	0	0	1(20)	0	0	0
4	10	1(10)	0	1(10)	1(10)	0	1(10)	0
4.5	5	4(80)	0	0	3(60)	1(20)	1(20)	1(20)
5	4	2(50)	0	1(25)	0	0	0	1(25)

Patient #19 was not evaluable for toxicities because of pretreatment ineligibilities (cardiac and renal failure). The only irreversible toxicities were hepatic failure in patient #28 and renal failure in patients #31 and #33. The renal failures in patients #31 and #33 were associated with infusion of high salt formulation DT$_{388}$GMCSF.

PTT, partial thromboplastin time.

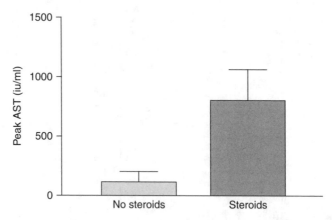

Figure 8.3 Bar graph showing mean peak AST (iu/ml) of DT$_{388}$GMCSF-treated patients with (799 ± 264, n = 6) or without (115 ± 87, n = 3) steroid prophylaxis. Only patients treated with ≥ 3 μg/kg/day and with measurable circulating DT$_{388}$GMCSF included. Difference was statistically significant (p = 0.03)

(p = 0.53) but was correlated with the pretreatment anti-DT antibody titer (p = 0.0001, see later). Interestingly, in the five patients for whom data were available, the peak DT$_{388}$GMCSF concentrations were higher on day 5 than on day 1.

The immune response to DT$_{388}$GMCSF was studied in all treated patients. We did not have a vaccination history for the 37 patients. However, most patients likely received their full immunization against DT in childhood. An enzyme-linked immunoassay (EIA) showed that 33 (90%) of 37 pretreatment patient sera were positive (i.e., had circulating anti-DT$_{388}$GMCSF antibodies) in concentrations ranging from 0.2 to 9.4 μg/ml with a median of 1.84 μg/ml.[64] Patients with low pretreatment antibody tiers were more likely to have measurable peak circulating DT$_{388}$GMCSF. The median and range of antibody for those with undetectable peak fusion protein during the first course were 2.5 μg/ml (1.1 to 9.4 μg/ml; n = 18), and the same antibody levels for those with measurable peak fusion protein were 0.6 μg/ml (0 to 3.7 μg/ml; n = 19). The difference was highly significant (p < 0.001). Only two of fourteen patients with EIA antibody concentrations above 2.2 μg/ml had detectable peak DT$_{388}$GMCSF compared with 17 of 23 with EIA antibody concentrations of < μg/ml. After 15 to 60 days, 22 of 26 evaluable patients had increased antibody titers ranging from 0.2 to 6.613 μg/ml. One patient had no change in antibody titer and three patients had decreased antibody titers. Of the three patients with no humoral immune response to DT$_{388}$GMCSF, one patient had received two prior allogeneic bone marrow grafts, one patient had undergone a prior autologous bone marrow transplant, and the last patient was heavily pretreated with fludarabine.

The DT fusion protein showed evidence of clinical efficacy. Four clinical remissions were observed. Six of the 37 patients had relapsed disease. Three of the four responses were seen in this group. Only one response occurred in a refractory patient. All the responders were treated with a combination of

DT$_{388}$GMCSF and steroids. Patient #22 was a 72-year-old female who developed AML in April 1996. She had normal cytogenetics and received idarubicin plus cytarabine (3 + 7), achieving a complete remission. She had no consolidation therapy, relapsed in January 2000, and received salvage therapy consisting of topotecan and high-dose cytarabine. However, she was refractory, and a bone marrow biopsy on March 8, 2000, showed 28% blasts confirmed by flow cytometry. She received a 5 day course of DT$_{388}$GMCSF at 5 µg/kg/day complicated only by asymptomatic and transient transaminasemia, elevated CPK, and hypocalcemia from April 10 to 14, 2000. Before therapy, she had an ANC of 280/µl, a platelet count of 64,000/µl, and no circulating blasts. One and two months posttherapy, her bone marrow showed 1 to 3% blasts by morphology and flow cytometry. She had recovery of platelets by day 60 to 158,000/µl but continued to be neutropenic (ANC of 279/µl). She was active spending most days out of the home and did not require antibiotics. By August 17, 2000, she had normalization of counts with an ANC of 1320/µl, a platelet count of 238,000/µl, and a hemoglobin of 12.0 gm/dl, and did not require transfusions nor antibiotics. A repeat bone marrow exam on November 15, 2000, showed no morphological evidence of increased blasts, but there were 8% blasts by flow cytometry. Her blood counts remained normal with an ANC of 1600/µl, a hemoglobin of 14.2 gm/dl, and a platelet count of 180,000/µl. By March 29, 2001, she had recurrence of pancytopenia (ANC of 420/µl, platelet count of 76,000/µl, and hemoglobin of 12.8 gm/dl). There were no circulating blasts, but the bone marrow examination showed 8% blasts by morphology and 12% blasts by flow cytometry. She received a second course of DT$_{388}$GMCSF at 5 µg/kg/day for 5 days, and the day 12 bone marrow showed disappearance of the blasts by morphology and flow cytometry. By day 17 after the second course, her ANC was 1000/µl. However, she remained thrombopenic (platelet count of 7000/µl) and anemic (hemoglobin of 7.1 gm/dl). On day 21, she developed a pneumonia, confirmed by chest X-ray, but declined aggressive management and died on day 24 posttherapy. Patient #24 was a 33-year-old female diagnosed with AML in November 1998 and induced with idarubicin plus cytarabine (3 + 7) followed by an allogeneic bone marrow transplant in June 1999. She developed bronchiolitis obliterans, requiring chronic corticosteroids, and had relapse of her AML in April 2000. Bone marrow exam showed 14% blasts. She had 300/µl circulating blasts and thrombopenia (platelet count of 23,000/µl). She received DT$_{388}$GMCSF at 5 µg/kg/day for 5 days, and again, the only side effects were transient, asymptomatic transaminasemia and hypocalcemia. Her day 30 bone marrow showed 2% blasts, but she remained pancytopenic with an ANC of 75/µl, a platelet count of 25,000/µl, and a hemoglobin of 10.8 gm/dl. She was asymptomatic. She received a course of granulocyte colony-stimulating factor (G-CSF) to stimulate recovery of normal myelopoiesis, however by day 60, a repeat bone marrow showed an increase in blast percentage to 15%. She was again treated with DT$_{388}$GMCSF at 5 µg/kg/day complicated only by asymptomatic and transient transaminasemia. Her marrow blasts at day 30 were 5 to 10%. She continued to have neutropenia and thrombopenia. She died 1 year after initiating fusion protein therapy in April 2001. Patient #25 was a 70-year-old male diagnosed with AML in March 2000 and treated with idarubicin plus cytarabine for induction and consolidation. He relapsed in April 2000 with bone marrow showing 15% blasts. He had circulating blasts (50/µl), thrombopenia (platelet count of 34,000/µl), and anemia (hemoglobin of 8.3 gm/dl). He was treated from June 22 to 26, 2000, with $DT_{388}GMCSF$ at 3 µg/kg/day. His course was complicated by transient renal insufficiency attributed to rofecoxib and transient, asymptomatic transaminasemia, hypocalcemia, and elevated CPK. The day 12 marrow showed 1% blasts, and the day 30 marrow showed 4% blasts. However, he remained anemic and thrombopenic. He declined further fusion protein or other therapy and died from progressive disease 40 days posttherapy. Patient #32 was a 19-year-old male diagnosed with AML in December 2000. His cytogenetics showed a t(8;21) translocation. He was given induction chemotherapy with idarubicin plus cytarabine (7 + 3) and achieved a remission. He had two consolidations with high-dose cytarabine, but soon relapsed in April 2000. He received fludarabine/cytarabine/G-CSF salvage but was unresponsive. He had 448/µl circulating blasts and his marrow showed 96% blasts. He was treated with $DT_{388}GMCSF$ 4 µg/kg/day with steroids from August 13 to 18, 2001. His course was complicated only by a asymptomatic transient grade 2 transaminase elevation. Day 15 bone marrow exam showed a reduction in blasts to 9% and there were no circulating blasts. The patient received an unrelated umbilical cord blood transplant on September 11, 2001, with conditioning consisting of total body irradiation, cyclophosphamide and ATG. He engrafted and follow-up marrow and blood showed no leukemia, but he had complications of transplantation including *candida krusei* bacteremia, Steven Johnson's syndrome, bladder hemorrhage with cystitis, seizure due to medications, CMV infection, aspergillus pneumonia and brain aspergillus, renal failure, grade 4 skin graft versus host disease, and expired on January 10, 2002. Autopsy showed no residual leukemia.

Future

We have initiated a new phase I study of $DT_{388}GMCSF$ in relapsed/refractory AML patients. The protocol, CCCWFU#27102, will (a) not give steroid prophylaxis, (b) select patients with low pretreatment anti-DT antibody titers (<2.4 µg/ml), and (c) treat Monday-Wednesday-Friday for two weeks to better monitor toxicities. The study has interpatient dose escalation with a starting dose of 4 µg/kg/day. Patients will receive only washed blood products to reduce anti-DT antibody exposure and receive prophylactic antibiotics (gatifloxacin and flucanozole), intravenous saline hydration, vitamin K, acetaminophen, and diphenhydramine. Since $DT_{388}GMCSF$ showed antileukemic activity with mild organ dysfunction in most patients, this additional study should be valuable in establishing whether this fusion protein may have a role in the care of these unfortunate patients.

Acknowledgments

We wish to acknowledge the support of the National Institutes of Health (R01CA76178, R01CA090263, and R21CA90550) and the Leukemia and Lymphoma Society (LSA6114).

References

1. Greenlee, RT, Hill-Harmon, MB, Murray, T, Thun, M. Cancer statistics, 2001. *CA Cancer J Clin* 51: 15–36, 2001.
2. Bennett, JM, Young, ML, Andersen, JW, Cassileth, PA, Tallman, MS, Paietta, E, Wiernik, PH, Rowe, JH. Long-term survival in acute myeloid leukemia: the Eastern Cooperative Oncology Group experience. *Cancer* 80 Suppl. 11: 2205–2209, 1997.
3. Vey, N, Keating, M, Giles, F, Cortes, J, Beran, M, Estey, E. Effect of complete remission on survival of patients with acute myelogenous leukemia receiving first salvage therapy. *Blood* 93: 3149–3150, 1999.
4. Estey, E. Treatment of relapsed and refractory acute myelogenous leukemia. *Leukemia* 14: 476–479, 2001.
5. Sievers, EL, Larson, RA, Stadtmauer, EA, Estey, E, Lowenberg, B, Dombret, H, Karanes, C, Theobald, M, Bennett, JM, Sherman, ML, Berger, MS, Eten, CB, Loken, MR, van Dongen, JJM, Bernstein, ID, Applebaum, FR. Efficacy and safety of Gemtuzumab Ozogamicin in patients with CD33-positive acute myeloid leukemia in first relapse. *J Clin Oncol* 19: 3244–3254, 2001.
6. Kern, W, Braess, J, Grote-Metke, A, Kuse, H, Fuchs, R, Hossfeld, DK, Reichle, A, Wormann, B, Buchner, T, Hiddemann, W. Combination of aclarubicin and etoposide for the treatment of advanced acute myeloid leukemia: results of a prospective multicenter phase II trial. *Leukemia* 12: 1522–1526, 1998.
7. Tsukaguchi, M, Furakawa, Y, Kitani, T. Low-dose etoposide and subcutaneous injection of low-dose cytosine arabinoside (Et-A non i.v.) with or without minimum anthracycline for refractory acute non-lymphoblastic leukemia. *Jap J Cancer Chemother* 25: 1603–1607, 1998.
8. Hurd, DD, Peterson, BA, McKenna, RW, Bloodfield, CD. VP16-213 and cyclophosphamide in the treatment of refractory acute nonlymphoblastic leukemia with monocytic features. *Med Ped Oncol* 9: 251–255, 1981.
9. Maraninchi, D, Vey, N, Viens, P, Stoppa, AM, Archimbaud, E, Attal, M, Baume, D, Bouabdallah, R, Demeoq, F, Fleury, J, Michallet, M, Olive, D, Rieffers, J, Sainty, D, Tabilio, A, Tiberghien, P, Brandely, M, Hercend, T, Blaise, D. A phase II study of interleukin-2 in 49 patients with relapsed or refractory acute leukemia. *Leuk Lymphoma* 31: 343–349, 1998.
10. Kantarjian, HM, Beran, M, Ellis, A, Zwelling, L, O'Brien, S, Cazenave, L, Koller, C, Rios, MB, Plunkett, W, Keating, MJ, Estey, EH. Phase I study of topotecan, a new topoisomerase I inhibitor, in patients with refractory or relapsed acute leukemia. *Blood* 81: 1146–1151, 1993.
11. Steins, MB, Padro, T, Bieker, R, Ruiz, S, Kropff, M, Kienast, J, Kessler, T, Buechner, T, Berdel, WE, Mesters, RM. Efficacy and safety of thalidomide in patients with acute myeloid leukemia. *Blood* 99: 834–839, 2002.
12. Seiter, K, Liu, D, Loughran, T, Siddiqui, A, Baskind, P, Ahmed, T. Phase I study of temozolomide in relapsed/refractory acute leukemia. *J Clin Oncol* 20: 3249–3253, 2002.

13. Jurcic, JG, Larson, SM, Sgouros, G, McDevitt, MR, Finn, RD, Divgi, CR, Ballangrud, AM, Hamacher, KA, Ma, D, Humm, JL, Brechbiel, MW, Molinet, R, Scheinberg, DA. Targeted alpha particle immunotherapy for myeloid leukemia. *Blood* 100: 1233–1239, 2002.

14. Giles, FJ. Toxacitabine-based therapy of refractory leukemia. *Expert Rev Anticancer Ther* 2: 261–266, 2002.

15. Gandhi, V, Plunkett, W, Du, M, Ayres, M, Estey, EH. Prolonged infusion of gemcitabine: clinical and pharmacodynamic studies during a phase I trial in relapsed acute myelogenous leukemia. *J Clin Oncol* 20: 665–673, 2002.

16. Lancet, JE, Rosenblatt, JD, Karp, JE. Farnesyltransferase inhibitors and myeloid leukemias: phase I evidence of Zarnestra activity in high-risk leukemias. *Semin Hematol* 39 Suppl. 2: 31–35, 2002.

17. Lowenberg, B, Sonneveld, P. Resistance to chemotherapy in acute leukemia. *Curr Opin Onc* 10: 31–35, 1998.

18. Gsur, A, Zochbauer, S, Gotzl, M, Kyrle, PA, Lechner, K, Pirker, R. MDR1 RNA expression as a prognostic factor in acute myeloid leukemia: an update. *Leuk Lymphoma* 12: 91–94, 1993.

19. List, A, Spier, C, Grogan, T, Johnson, C, Roe, D, Greer, J, Wolff, S, Broxterman, H, Scheffer, G, Scheper, R, Dalton, W. Overexpression of the major vault transporter protein lung-resistance protein predicts treatment outcome in acute myeloid leukemia. *Blood* 87: 2464–2469, 1996.

20. Russo, D, Marie, JP, Zhou, DC, Faussat, AM, Melli, C, Damiani, D, Micheleutti, A, Michieli, M, Fanin, R, Baccarani, M. Evaluation of the clinical relevance of the anionic glutathione-S-transferase (GSTπ) and multi-drug resistance z (mdr-1) gene co-expression in leukemias and lymphomas. *Leuk Lymphoma* 15: 453–468, 1994.

21. Fenaux, P, Preudhomme, C, Quiquandon, I, Jonveaux, P, Lai, JL, Vamrumbeke, M, Loucheux-Lefebvre, MH, Bauters, F, Berger, R, Kerckaert, JP. Mutations of the p53 gene in acute myeloid leukemia. *Br J Haematol* 80: 178–183, 1992.

22. Campos, L, Rouault, J, Sabido, O, Oriol, P, Roubi, N, Vasselon, C, Archimbaud, E, Magaud, J, Guyotat, D. High expression of bcl-2 protein in acute myeloid leukemia cells. *Blood* 77: 2404–2412, 1991.

23. Steinbach, D, Sell, W, Voigt, A, Hermann, J, Zintl, F, Sauerbrey, A. BCRP gene expression is associated with a poor response to remission induction therapy in childhood acute myeloid leukemia. *Leukemia* 16: 1443–1447, 2002.

24. Galmarini, CM, Thomas, X, Calvo, F, Rousselot, P, Jafaari, AE, Cros, E, Dumontet, C. Potential mechanisms of resistance to cytarabine in AML patients. *Leuk Res* 26: 621–629, 2002.

25. Sauerbrey, A, Zintl, F, Hermann, J, Volm, M. Multiple resistance mechanisms in acute nonlymphoblastic leukemia (ANLL). *Anticancer Res* 18: 1231–1236, 1998.

26. Myers, DE, Uckun, FM, Ball, ED, Vallera, DA. Immunotoxins for *ex vivo* marrow purging in autologous bone marrow transplantation for acute nonlymphocytic leukemia. *Transplantation* 46: 240–245, 1988.

27. Roy, DC, Griffin, JD, Belvin, M, Blattler, WA, Lambert, JM, Ritz, J. Anti-MY9-blocked ricin: an immunotoxin for selective targeting of acute myeloid leukemia cells. *Blood* 77: 2404–2412, 1991.

28. McGraw, KJ, Rosenblum, MG, Cheung, L, Scheinberg, DA. Characterization of murine and humanized anti-CD33, gelonin immunotoxins reactive against myeloid leukemias. *Cancer Immunol Immunother* 39: 367–374, 1994.

29. Benedetti, G, Bondesan, P, Caracciolo, D, Cherasco, C, Ruggieri, D, Gastaldi, M, Pileri, A, Gianni, A, Tarella, C. Selection and characterization of early hematopoietic progenitors using an anti-CD71/SO6 immunotoxin. *Exp Hematol* 22: 166–173, 1994.

30. Burbage, C, Tagge, EP, Harris, B, Hall, P, Fu, T, Willingham, MC, Frankel, AE. Ricin fusion toxin targeted to the human granulocyte-macrophage colony-stimulating factor is selectively toxic to acute myeloid leukemia cells. *Leuk Res* 21: 681–690, 1997.

31. Kreitman, RJ, Pastan, I. Recombinant toxins containing human granulocyte-macrophage colony-stimulating factor and either Pseudomonas exotoxin or diphtheria toxin kill gastrointestinal cancer and leukemia cells. *Blood* 90: 252–259, 1997.

32. Tagge, E, Harris, B, Burbage, C, Hall, P, Vesely, J, Willingham, M, Frankel, A. Synthesis of green fluorescent protein-ricin and monitoring of its intracellular trafficking. *Bioconj Chem* 8: 743–750, 1997.

33. Choe, S, Bennett, MJ, Fujii, G, Curmi, PMG, Kantardjieff, KA, Collier, RJ, Eisenberg, D. The crystal structure of diphtheria toxin. *Nature* 357: 216–222, 1992.

34. Cha, J, Brooke, JS, Ivey, KN, Eidels, L. Cell surface monkey CD9 antigen is a coreceptor that increases diphtheria toxin sensitivity and diphtheria toxin receptor affinity. *J Biol Chem* 275: 6901–6907, 2000.

35. Simpson, JC, Smith, DC, Roberts, LM, Lord, JM. Expression of mutant dynamin protects cells against diphtheria toxin but not against ricin. *Exp Cell Res* 239: 293–300, 1998.

36. Rosconi, MP, London, E. Topography of helices 5-7 in membrane-inserted diphtheria toxin T domain: indentification and insertion boundaries of two hydrophobic sequences that do not form a stable transmembrane hairpin. *J Biol Chem* 277: 16517–16527, 2002.

37. Han, S, Tainer, JA. The ARTT motif and a unified structural understanding of substrate recognition in ADP-ribosylating bacterial toxins and eukaryotic ADP-ribosyltransferases. *Int J Med Microbiol* 291: 523–529, 2002.

38. Thorburn, J, Frankel, AE, Thorburn, A. Apoptosis by leukemia cell-targeted diphtheria toxin occurs via receptor-independent activation of Fas associated death domain protein. *Clin Cancer Res*, in press.

39. Frankel, AE, Rossi, P, Kuzel, TM, Foss, F. Diphtheria fusion protein therapy of chemoresistant malignancies. *Curr Cancer Drug Targets* 2: 19–36, 2002.

40. Lilly, MB, Frankel, AE, Salo, J, Kraft, AS. Distinct domains of the human GM-CSF receptor alpha subunit mediate activation of JAK/STAT signaling and differentiation. *Blood* 97: 1662–1670, 2001.

41. Rozwarski, DA, Diederichs, K, Hecht, R, Boone, T, Karplus, PA. Refined crystal structure and mutagenesis of human granulocyte-macrophage colony-stimulating factor. *Proteins* 26: 304–313, 1996.

42. Park, LS, Waldron, PE, Friend, D, Sassenfeld, HM, Price, V, Anderson, D, Cosman, D, Andrews, RG, Bernstein, ID, Urdal, DL. Interleukin-3, GM-CSF, and G-CSF receptor expression on cell lines and primary leukemia cells: receptor heterogeneity and relationship to growth factor responsiveness. *Blood* 74: 56–65, 1989.

43. Cannistra, SA, Koenigsmann, M, DiCarlo, J, Groshek, P, Griffin, JD. Differentiation-associated expression of two functionally distinct classes of granulocyte-macrophage colony-stimulating factor receptors by human myeloid cells. *J Biol Chem* 265: 12656–12663, 1990.

44. Frankel, AE, Ramage, J, Latimer, A, Feely, T, Delatte, S, Hall, P, Tagge, E, Kreitman, R, Willingham, M. High-level expression and purification of the recombinant diphtheria fusion toxin DTGM for PHASE I clinical trials. *Protein Expression Purification* 16: 190–201, 1999.
45. Frankel, AE, Hall, PD, Burbage, C, Vesely, J, Willingham, M, Bhalla, K, Kreitman, RJ. Modulation of the apoptotic response of human myeloid leukemia cells to a diphtheria toxin-granulocyte-macrophage colony stimulating factor (GMCSF) fusion protein. *Blood* 90: 3654–3661, 1997.
46. Hogge, DE, Willman, CL, Kreitman, RJ, Berger, M, Hall, PD, Kopecky, KJ, McLain, C, Tagge, EP, Eaves, CJ, Frankel, AE. Malignant progenitors from patients with acute myelogenous leukemia are sensitive to a diphtheria toxin-granulocyte-macrophage colony stimulating factor (GMCSF) fusion protein. *Blood* 92: 589–595, 1998.
47. Frankel, AE, Lilly, M, Kreitman, RJ, Hogge, D, Beran, M, Freedman, MH, Emanuel, PD, McLain, C, Hall, P, Tagge, EP, Berger, M, Eaves, C. Diphtheria toxin fused to granulocyte-macrophage colony-stimulating factor is toxic to blasts from patients with juvenile myelomonocytic leukemia and chronic myelomonocytic leukemia. *Blood* 92: 4279–4286, 1998.
48. Feuring-Buske, M, Frankel, AE, Gerhard, B, Hogge, D. Cytotoxicity of diphtheria toxin 388-granulocyte macrophage colony-stimulating factor (DT388-GMCSF) fusion protein for acute myelogenous leukemia (AML) stem cells. *Exp Hematol* 28: 1390–1400, 2000.
49. Parentesis, JP, Waddick, KG, Bendel, AE, Shao, Y, Warman, BE, Chandan-Langlie, M, Uckun, FM. Induction of apoptosis in multidrug-resistant and radiation-resistant acute myeloid leukemia cells by a recombinant fusion toxin directed against the human granulocyte-macrophage colony-stimulating factor receptor. *Clin Cancer Res* 3: 347–355, 1997.
50. Rozemuller, H, Rombouts, EJC, Touw, IP, FitzGerald, DJP, Kreitman, RJ, Pastan, I, Hagenbeek, A, Martens, ACM. Sensitivity of human acute myeloid leukemia to diphtheria toxin-GM-CSF fusion protein. *Br J Haematol* 98: 952–959, 1997.
51. Perentesis, JP, Bendel, AE, Shao, Y, Warman, B, Yang, CH, Chandan-Langlie, M, Waddick, KG, Uckun, FM. Granulocyte-macrophage colony-stimulating factor receptor-targeted therapy of chemotherapy- and radiation-resistant human myeloid leukemias. *Leuk Lymphoma* 25: 247–256, 1997.
52. Bendel, AE, Shao, Y, Davies, SM, Warman, B, Yang, CH, Waddick, KG, Uckun, FM, Perentesis, JP. A recombinant fusion toxin targeted to the granulocyte-macrophage colony-stimulating factor receptor. *Leuk Lymphoma* 25: 257–270, 1997.
53. Terpstra, W, Rozemuller, H, Breems, DA, Rombouts, EJ, Prins, A, FitzGerald, DJ, Kreitman, RJ, Wielenga, JJ, Ploemacher, RE, Lowenberg, B, Hagenbeek, A, Martens, AC. Diphtheria toxin fused to granulocyte-macrophage colony-stimulating factor eliminates acute myeloid leukemia cells with the potential to initiate leukemia in immunodeficient mice, but spares normal hematopoietic stem cells. *Blood* 90: 3735–3742, 1997.
54. Frankel, AE, Hall, PD, McLain, C, Safa, AR, Tagge, EP, Kreitman, RJ. Cell-specific modulation of drug resistance in acute myeloid leukemic blasts by targeted toxins. *Bioconj Chemistry* 9: 490–496, 1998.
55. Kim, CN, Bhalla, K, Kreitman, RJ, Willingham, MC, Hall, PD, Tagge, EP, Jia, T, Frankel, AE. Diphtheria toxin fused to granulocyte-macrophage

colony-stimulating factor and ara-C exert synergistic toxicity against human AML HL-60 cells. *Leuk Res* 23: 527–538, 1999.

56. Thorburn, J, Frankel, AE, Thorburn, A. Apoptosis by leukemia cell-targeted diphtheria toxin occurs via receptor-independent activation of Fas associated death domain protein. *Clin Cancer Res*, in press.

57. Hall, PD, Kreitman, RJ, Willingham, MC, Frankel, AE. Toxicology and pharmacokinetics of DT388-GMCSF, a fusion toxin consisting of a truncated diphtheria toxin (DT388) linked to human granulocyte-macrophage colony stimulating factor (GMCSF), in C57Bl/6 mice. *Tox Appl Pharm* 150: 91–97, 1998.

58. Hall, PD, Kreitman, RJ, Willingham, MC, Tagge, EP, Frankel, AE. Diphtheria toxin fused to granulocyte-macrophage colony-stimulating factor prolongs the survival of SCID mice bearing human acute myelogenous leukemia. *Leukemia* 13: 629–633, 1999.

59. Perentesis, JP, Gunther, R, Waurzyniak, B, Yanishevski, Y, Myers, DE, Ek, O, Messinger, Y, Shao, Y, Chelstrom, LM, Schneider, E, Evans, WE, Uckun, FM. *In vivo* biotherapy of HL-60 myeloid leukemia with a genetically engineered recombinant fusion toxin directed against the human granulocyte-macrophage colony-stimulating factor receptor. *Clin Cancer Res* 3: 2217–2227, 1997.

60. Rozemuller, H, Terpstra, W, Rombouts, EJC, Lawler, M, Byrne, C, FitzGerald, DJP, Kreitman, RJ, Wielenga, JJ, Lowenberg, B, Touw, IP, Hagenbeek, A, Martens, ACM. GM-CSF receptor targeted treatment of primary AML in SCID mice using diphtheria toxin fused to huGM-CSF. *Leukemia* 12: 1962–1970, 1998.

61. Hotchkiss, CE, Hall, PD, Cline, JM, Willingham, MC, Kreitman, RJ, Gardin, J, Latimer, A, Ramage, J, Feely, T, Frankel, AE. Toxicology and pharmacokinetics of DTGM, a fusion toxin consisting of a truncated diphtheria toxin (DT388) linked to human granulocyte-macrophage colony-stimulating factor (GM-CSF), in cynomolgus monkeys. *Tox Appl Pharm* 158: 152–160, 1999.

62. Rozemuller, H, Rombouts, EJC, Touw, IP, FitzGerald, DJ, Kreitman, RJ, Hagenbeek, A, Martens, C. *In vivo* targeting of leukemic cells using diphtheria toxin fused to murine GM-CSF. *Leukemia* 12: 710–717, 1998.

63. Frankel, AE, Powell, BL, Hall, PD, Case, LD, Kreitman, RJ. Phase I trial of a novel diphtheria toxin/granulocyte-macrophage colony-stimulating factor fusion protein (DT$_{388}$GMCSF) for refractory or relapsed acute myeloid leukemia. *Clin Cancer Res* 8: 1004–1013, 2002.

64. Hall, PD, Virella, G, Willoughby, T, Atchley, DH, Kreitman, RJ, Frankel, AE. Antibody response to DT-GM, a novel fusion toxin consisting of a truncated diphtheria toxin (DT) linked to human granulocyte-macrophage colony-stimulating factor (GM), during a phase I trial of patients with relapsed or refractory acute myeloid leukemia. *Clin Immunol* 100: 191–197, 2001.

chapter nine

VEGF-Diphtheria toxin for targeting tumor blood vessels

Indira V. Subramanian and Sundaram Ramakrishnan

Contents

Introduction

The formation of new blood vessels by either angiogenesis or vasculogenesis is an essential step in embryonic development and in the etiology of many diseases, including cancer. Vasculogenesis has been implicated as a major contributor to tumor vascularization because up to 40% of the endothelial cells in a tumor was found to be derived from endothelial progenitor cells that originate in the bone marrow.[1] Vascular endothelial growth factor (VEGF) is the most critical inducer of blood vessel formation. VEGF initiates the formation of immature vessels by vasculogenesis or angiogenic sprouting during embryonic development as well as in adults. The morphology and functions of endothelial cells vary extensively among different organs. While endothelial lineage is genetically determined, expression of a set of vascular bed genes is regulated at the transcriptional level following interaction with local signaling pathways.[2] The VEGF family of proteins includes VEGF-A, -B,-C,-D,-E, and placental growth factor (PlGF). VEGF-A, also called VPF, vascular permeability factor, has been considered one of the most important positive regulators of tumor angiogenesis.[3] Human VEGF is secreted as a disulfide-linked dimeric glycoprotein with five different isoforms (VEGF206, 189, 165, 145, and 121) produced by means of alternate splicing and has unique biological functions.[4] Mouse VEGF is highly homologous to human VEGF but is shorter by a single amino acid residue. Interestingly, antibodies generated against human VEGF cross-react and neutralize the biological activity of VEGF from rodents. VEGF binds to three types of receptors, flt-1 (VEGFR-1), KDR/flk-1 (VEGFR-2), and flt-4 (VEGFR-3). All three receptors consist of seven Ig-like domains, a transmembrane sequence, and an intra-cellular kinase domain (Figure 9.1).[5-7] VEGF binds preferentially to VEGFR-1 and VEGFR-2, whereas VEGF-C and VEGF-D interact with VEGFR-3. Additionally, VEGF165 binds to neuropilin (NP-1) via the heparin-binding domain encoded by the exon-7. Although NP-1 does not directly transduce signaling, it is an integral part of the VEGF165–VEGFR complex. NP-1 may also sequester VEGF and present it to cognate receptors.[8] All members

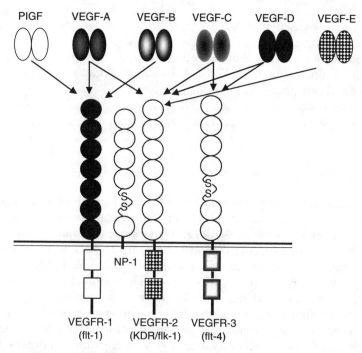

Figure 9.1 Vascular endothelial growth factor (VEGF) families of proteins are key mediators of angiogenesis. Individual members of this family have many splice variants as well. One of the splice variants of VEGF-A is VEGF121, which lacks a heparin binding domain encoded by the exon 7. A splice variant of VEGF-A containing a heparin binding domain interacts with the co-receptor, neuropilin-1 (NP-1)

of VEGF bind to cell-surface receptors (VEGFRs), which lead to ligand-dependent receptor dimerization and activation of the receptor tyrosine kinase to initiate a cellular response.[9] Cocrystallization studies showed that the second Ig-domain of the receptors makes direct contact with VEGF molecule. The role of VEGFR-1 is less clear, but it might mediate cell migration and vascular permeability. VEGFR-2 is the key receptor that triggers endothelial cell proliferation, angiogenesis, and vascular permeability.[10,11] Upon receptor activation, a soluble component of the extracellular domain of the VEGFR is observed in the culture media, which is believed to attenuate the signaling cascade. Soluble receptors are also seen in biological fluids. Soluble receptors have been cloned as fusion proteins with Fc and are used to inhibit angiogenesis *in vivo*.

Receptors for VEGF (VEGFR-1 and VEGFR-2) with the exception of NP-1 are expressed almost exclusively on endothelial cells and they are markedly overexpressed in the stromal vessels supplying tumors.[9] Normal quiescent blood vessels show negligible levels of VEGFR-1 and VEGFR-2. In fact, *in situ* hybridization studies have shown overexpression of VEGFR-1 and VEGFR-2 in vasculatures inside the tumors and in the peritumoral areas.[12] Transcripts for VEGFR were almost undetectable in the adjoining normal tissues. Moreover, tumor-associated vascular endothelium proliferates at a faster rate than the quiescent endothelium and is therefore vulnerable to antiproliferative endothelial cell-targeted therapy.[13] High levels of VEGFR expression in the tumor vasculature thus provide a unique opportunity for tumor targeting.[14]

Recently, a tissue-specific angiogenic factor, endocrine gland vascular endothelial growth factor (EG-VEGF), has been identified.[15] Expression of human EG-VEGF mRNA is restricted to the steroidogenic glands, ovary, testis, adrenal, and placenta. EG-VEGF promoted proliferation, migration, and fenestration in cultured capillary endothelial cells derived from such endocrine glands. It also induced extensive angiogenesis when delivered in the ovary but not in the cornea. In contrast, VEGF induced angiogenesis in all the organs tested. EG-VEGF binds to a distinct class of receptor containing seven transmembrane regions, which is a characteristic of G-protein coupled receptors. We are yet to learn more about the distribution of this receptor in tumor tissues and whether this receptor could be used as a target for delivering toxin polypeptides. Other new targets in the tumor endothelium have been identified by comparing libraries of genes (e.g., SAGE, serial analysis of gene expression, libraries) constructed from isolated endothelial cells from tumors and normal tissues.[16] This approach has identified three novel, transmembrane, endothelial-specific genes called TEM1, TEM5, and TEM8.[17] Several other endothelial-specific genes have been identified by using alternate strategies. Identification of *robo4* as an endothelial-specific gene in tumors is of considerable interest.[18] The *robo4* gene is present on the cell surface and is ideal for targeting. Moreover it is a developmental gene and it is not expressed in adult tissue. Another such gene is *Delta4*, which is also endothelial specific and again is found only on tumor endothelium.[19]

Targeting tumor vasculature

Most of the currently available cancer treatment strategies attempt to target cancer cells directly. These methods took advantage of biochemical and immunological differences between normal and neoplastic cells in selectively inhibiting tumor growth. While these methods are still valid and clinically useful, a complementary targeting approach could pontentiate the effects of conventional therapies. Vascular endothelium is emerging as a possible candidate. There are many advantages to this strategy: (A) Vascular endothelial cells constitute a small fraction of the total tumor mass. One endothelial cell is seen for every 100 to 500 tumor cells. Therefore, only a small number of endothelial cells have to be eliminated to inhibit a larger tumor mass. (B) Endothelial cells are normal cells and, therefore, do not exhibit drug resistance. (C) Endothelial cells from normal tissues are quiescent and multiply once in 6 months. Tumor vascular endothelial cells on the other hand rapidly divide to meet the growth demands of tumor tissue. Such differences provide a unique opportunity to selectively inhibit tumor neovascularization. Vascular targeting strategies can be divided into two categories, those using biological agents and those using small molecules as drugs. The biological agents involve ligand-directed vascular targeting, which uses antibodies, peptides, toxins, procoagulants, and proapoptotic effectors to tumor endothelium. The small molecules do not specifically localize to tumor endothelium, but exploit pathophysiological differences between tumor and normal tissue endothelia to induce selective effects on tumor vessels.[20] In this chapter we focus on the effects of VEGF–toxin conjugates on tumor angiogenesis.

VEGF–toxin conjugates

VEGF stimulates endothelial cell proliferation and migration by binding to two distinct cell-surface receptor tyrosine kinases. Receptor activation induces tyrosine phosphorylation of cytoplasmic signaling proteins that contain SH2 domains.[21] These proteins allow KDR/flk-1 and Flt-1 to communicate with the signaling pathways that promote responses to VEGF. Cell-surface receptors undergo endocytosis after the formation of a ligand–receptor complex. An acidic environment in endosomes promotes dissociation of ligand–receptor complexes. Subsequently, the ligand, the receptor, or both can recycle to the cell surface or are routed to the lysosomes for degradation.[22–24] Therefore substituting the binding domain of the toxins with VEGF can direct the toxin molecules to the intracellular compartments of endothelial cells. The intracellular processing of ligand-bound VEGFR is not completely understood. Proof of principle for this approach had been obtained by the chemical linking of VEGF with a truncated diphtheria toxin (DT) moiety.

DT is secreted as a mature protein of 535 residues with an M_r of 58,342.[25] DT is proteolytically processed (nicked) into two fragments, DTA fragment (193 amino acid residues) and DTB fragment (342 residues) linked by a disulfide bond between Cys 186 and Cys 201. The DTB fragment contains binding domain at the carboxyl terminus and four hydrophobic regions, which help in the transfer of the DTA fragment into the cytoplasm (functionally equivalent to the translocation domain of *Pseudomonas* exotoxin A). The DTA fragment ADP, ribosylates a posttranslationally modified histidine residue, diphthamide, present in elongation factor 2 to inhibit protein synthesis in mammalian cells.[26,27] The DT is truncated at residue 385 (DT385). DT385 contains the catalytic domain and the putative translocation domain but lacks the innate receptor-binding domain (Figure 9.2). Truncated DT has been previously found to be nontoxic to human cells but useful in generating effective fusion proteins for targeting. DT385 was further genetically modified to incorporate a cystine residue at the carboxyl terminus (DT385–Cys) to facilitate chemical conjugation to VEGF.[28] In our lab DT385 was expressed in *E. coli* and purified using a Nickel-NTA affinity column followed by gel filtration.[29] VEGF165 and VEGF121 were expressed in *Pichia pastoris* and were then chemically conjugated to the DT-385 mutant containing a carboxyl terminus cystine residue. VEGF165 and VEGF121 were derivatized with the heterobifunctional agent *N*-succinimidyl-3-(2-pyridyldithio) propionate (SPDP) to introduce 1–3 activated thiol groups by modifying the ε-amino group of lysine residues.[28] Derivatized VEGF was then incubated with DT385-Cys to establish a disulfide bond between the toxin moiety and VEGF by thiol exchange. VEGF–DT385 conjugate was then purified by affinity chromatography and gel filtration.

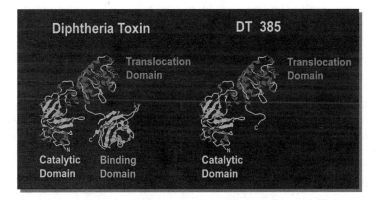

Figure 9.2 The diphtheria toxin has three structurally and functionally distinct domains. The amino-terminus domain is the site of catalytic activity. The carboxy-terminus has a receptor binding domain. Between the catalytic and receptor binding domain is a putative translocation domain, which undergoes conformational changes under acidic pH of endosomal vesicles. The translocation domain facilitates the cytoplasmic transport of the catalytic domain. DT385 was generated by the genetic deletion of the receptor binding domain from CRM-107. Additionally, DT385 was mutated to incorporate a cys residue at the C-terminus. Free thiol group of the C-terminal cys was used to chemically link DT385 to VEGF. Structural data from M.J. Bennette and D. Eisenberg (Protein Data Bank # 1MDT) was used to generate a truncated version of DT by the Swiss-PD view program

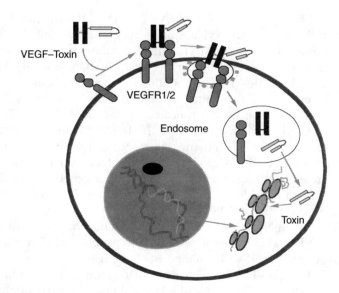

Figure 9.3 Schematic diagram showing receptor mediated endocytosis of the VEGF–toxin conjugate

The biological activity of VEGF165 is modulated by the heparin-binding domain. Through this region, VEGF165 can also interact with NP-1. The VEGF121 does not bind to NP-1. In addition, VEGF121 binds selectively to VEGFR-2 compared to VEGFR-1. Targeting toxin polypeptides by VEGF165 would bind to VEGFR-1 and VEGFR-2 on the tumor vessels as well as NP-1 on certain normal tissues. VEGFR-1 binds VEGF-A with a K_d of approximately 10 pM (7). Despite this high affinity binding, the effect of VEGF-A treatment on VEGFR-1 kinase activity is poor. The binding affinity of VEGF-A for VEGFR-2 is lower (K_d 75–125 pM) than for VEGFR-1 (K_d 10–25 pM), but biological activities of VEGF-A are thought to be transduced mainly by VEGFR-2.[30] VEGFRs are expressed most abundantly in the endothelium of sprouting blood vessels. Expression of both VEGF and its receptors are under the transcriptional regulation of hypoxia inducible factor 1-alpha (HIF1α). Toxin conjugates are internalized by receptor-mediated endocytosis into the endocytic vesicles and undergo pH-dependent conformational changes leading to translocation of the toxin moiety into the cytosol (Figure 9.3). Once in the cytoplasmic compartment the catalytic fragment of DT inhibits protein synthesis by adenosine diphosphate (ADP)-ribosylation of elongation factor 2.[28,31]

Inhibition of endothelial cell proliferation and angiogenesis

In vitro cytotoxicity assays show that both VEGF165 and VEGF121–DT385 are equally effective in inhibiting endothelial cell proliferation. Data in

Figure 9.4(A) show the dose-dependent inhibition of human umbilical vein endothelial cell (HUVEC) proliferation. Both VEGF165 and VEGF121–DT385 conjugates inhibited completely at 100 nM concentrations.[32,33] Additional binding of VEGF165–DT385 to NP-1 does not seem to alter the extent of cytotoxicity. Conjugates selectively inhibited endothelial cells and not receptor negative fibroblasts or tumor cells. Interestingly, the effect of VEGF–toxin conjugate on endothelial cells appears to be dependent on the proliferation status. Proliferating, subconfluent cultures were selectively inhibited by VEGF–toxin conjugate. Confluent cultures are G_0 arrested and are completely resistant to the action of VEGF–toxin (Figure 9.4(B)). Differences in sensitivity between proliferating and quiescent endothelial cells can be attributed to (a) variations in receptor-mediated endocytosis or (b) differential intracellular routing and compartmentalization. Resistance of quiescent

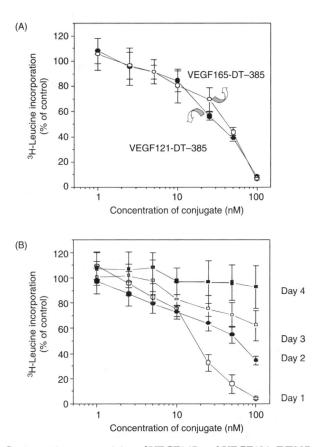

Figure 9.4 Comparative cytotoxicity of VEGF165 and VEGF121–DT385 conjugate on human umbilical vein endothelial cells. Panel A shows inhibition of proliferation in the presence of VEGF–toxin conjugates. Panel B shows the cytotoxic effect on proliferating and quiescent endothelial cells. Confluent cultures are relatively resistant to the VEGF–toxin conjugate

endothelial cells to VEGF–toxin conjugate can be advantageous *in vivo* and could help in the selective killing of tumor vascular endothelial cells. Endothelial cells of tumor vessels are highly proliferative and the normal tissues are quiescent.

The effect of VEGF–toxin conjugates to inhibit angiogenesis was evaluated in different model systems. Addition of conjugate completely inhibited the tube formation in matrigels. Application of conjugate-coated cover slips inhibited developmental angiogenesis in chick chorioallantoic membrane (CAM) assays. Parenteral administration of VEGF–toxin conjugate was also effective in inhibiting tumor cell-induced angiogenesis of matrigel plugs subcutaneously implanted into athymic nude mice. In these studies, LS 174 colon carcinoma cells were injected subcutaneously into the flanks of athymic nude mice. Two days after implantation, animals were randomized and treated with VEGF–toxin conjugate. Mice received a daily dose of 20 μg VEGF165–DT385 toxin conjugate or saline control for a 7-day period. At this stage, animals were sacrificed and tumor specimens were taken for histological analysis. Cryocut sections were stained with a PE-conjugated antibody against mouse CD-31, a tumor vessel marker.[34] Computer assisted image analysis was used to determine microvessel density and tumor vessel architecture (Figure 9.5). Mice treated with VEGF–DT conjugate showed 41.3% decrease in total vessel length, which was statistically significant ($p < 0.002$) when compared to the control group of mice that were treated with saline.

Inhibition of tumor growth by VEGF–DT toxin conjugates

Both VEGF165 and VEGF121 conjugated to DT385 were effective in inhibiting tumor growth. Two xenotransplant models were used to evaluate the effect of VEGF–toxin conjugates. Human ovarian cancer cell line, MA 148, was injected subcutaneously to establish tumors. After 7 days the conjugate was administered i.p. at a dose of 10 μg/day/mouse. After 5 weeks the average tumor volume in the control group of mice injected with saline was 2742 mm^3 whereas the tumor volume in mice administered with VEGF165–DT385 toxin conjugate was 820 mm^3 ($p < 0.02$). Histology and gross examination of tissues such as the lung, liver, and kidney showed no symptoms of tissue injury, in particular, hemorrhage, inflammatory infiltrates, or necrosis. In the same animals, however, the tumor tissues showed significant hemorrhage and necrosis. These studies demonstrated that the conjugate selectively inhibited the tumor vasculature.[32] Similarly, VEGF121–DT385 conjugate treatment significantly inhibited the growth of C6 rat glioma cells transplanted into athymic nude mice. VEGF121–DT385 conjugate treatment reduced the size of the tumors by threefold when compared to the control group (Figure 9.6). Histological analyses revealed that there was a reduction in vascularity in the residual tumors resected from the mice. Concomitant with the reduction in

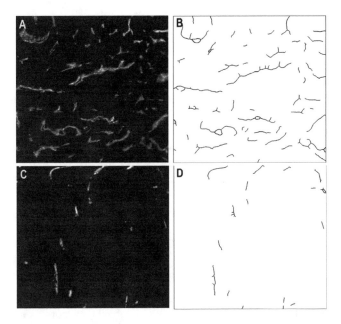

Figure 9.5 Human colon cancer cell line, LS174T was resuspended in matrigel and implanted subcutaneously into the flanks of athymic nude mice. The mice were treated with daily injections of either PBS or VEGF–DT385 conjugate. After 1 week matrigel plugs were surgically removed and snap frozen. Sections of matrigel plugs were stained with anti-mouse CD31–PE conjugate. Images were then "skeletonized" according to the method of Wild et al.[34] Panels A and B: representative images from control mice showing higher number of blood vessels. Panels C and D: representative images from the VEGF-DT385 conjugate treated mice showing reduced number of blood vessels

vessel density, there was an increase in the number of apoptotic tumor cells. TUNEL assays showed an eight- to ninefold increase in the apoptotic index ($p < 0.008$) in the tumors treated with the conjugate as compared to the control tumors.[33] The effects of VEGF165–DT385 conjugate treatment was also investigated on spontaneous tumorigenesis of mammary adenocarcinomas in C3 (1)/Tag transgenic mice. These studies showed that conjugate treatment during the period of angiogenic switch clearly delayed the appearance of tumors and number of lesions per mouse (Wild et al. unpublished data).

As an alternate to chemical conjugation methods other investigators have constructed DT–VEGF fusion proteins by recombinant methods.[31] Fusion protein, DT390–VEGF165 and DT390–VEGF121 were effective against Kaposi's sarcoma *in vitro* and *in vivo*. Tumor growth was inhibited in mice receiving doses of either fusion toxin when compared to control mice treated with vehicle alone.[31] In addition to truncated DT, other toxin moieties have also been used to prepare fusion proteins to target endothelial cells.

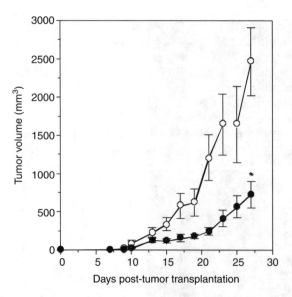

Figure 9.6 Effect of VEGF–toxin conjugate on tumor growth. Athymic nude mice were transplanted subcutaneously with C6 glioma cells. Tumors were allowed to establish for 7 days. Then, groups of mice were treated with 10 μg of the VEGF121–DT385 conjugate for a period of 14 days. Conjugate treatment significantly inhibited the growth of gliomas

The catalytic A-subunit of Shiga-like toxin I was fused to VEGF121 (SLT–VEGF). SLT-1 produced by *E. coli* O157:H7 is a "natural killer" for endothelial cells. Damage to endothelial cells caused by SLT-1 leads to the pathogenesis of hemorrhagic colitis and hemolytic uremic syndrome induced by *E. coli*.[35] Studies indicate that SLT–VEGF inhibits growing cells overexpressing VEGFR-2.[36] It was reported that the number of VEGFR-2 receptors on endothelial cells determines their sensitivity to SLT–VEGF fusion protein. Fusion protein induced apoptosis without significant inhibition of protein synthesis.[37] Gelonin (rGel) is a single chain N-glycosidase similar in action to Ricin A Chain.[38] The VEGF121 fusion toxin containing Gelonin was localized in tumor blood vessels after i.v. administration. Vascular damage and thrombosis of tumor blood vessels were observed within 48 h of administration of VEGF121–rGel to prostrate carcinoma xenografts, consistent with the primary action of the construct on tumor vasculature.[39] In conclusion, VEGF–toxin conjugates made either by chemical linkage or by genetic fusion are effective reagents capable of inhibiting angiogenesis and tumor growth.

Problems and prospects

One of the limitations of VEGF–toxin conjugate is that it can target and kill non-endothelial cells expressing VEGFR as well. DT390–VEGF165 fusion protein for example was found to be cytotoxic to the retinal pigment epithelial cells (RPE) *in vitro* whereas the unconjugated DT did not have any effect on these cells. RPE cell counts were decreased by 44% and by 56% after 3 days of treatment with 50 and 100 ng/ml of conjugate respectively.[40] Compartmentalized delivery and pharmacological modulation of conjugate administration can eliminate some of these problems. Another potential limitation for further clinical development of VEGF–toxin conjugate is the immune response against the toxin moiety that is likely to occur in patients. Antibodies to toxins will limit the duration of treatment and therefore will reduce efficacy. One possible way to overcome this problem is to use effector proteins (e.g., caspase activating enzymes) derived from human. The RNase A superfamily is also a good substitute for bacterial toxins. The RNase A family consists of both tissue-specific and ubiquitously expressed enzymes that are potentially toxic to cells. The active sites of RNase are blocked by ribonuclease inhibitor (RI) and protect the cellular RNA from RNases. RNases that elude RI are currently in clinical trials for the treatment of cancer.[41] In general, RNases are not specifically taken up by tumors, and therefore conjugation of RNases to targeting ligands improves their distribution and toxicity profile. In order to test the potential of RNase conjugates as target-directed toxins in cancer therapy, Suzuki et al.[42] conjugated human RNase A to two different proteins known to bind to transferrin receptors, which are overexpressed in cancer cells. When either transferrin or a monoclonal antibody directed against the transferrin receptor was conjugated to human RNase A, the resulting RNase conjugates became resistant to inhibition by RI, and exhibited cytotoxicity toward cell lines expressing transferrin receptors. More recently, targeting of apoptotic machinery is gaining importance. Most of the cytotoxic agents are now thought to kill cells predominantly by triggering apoptotic pathways.[43,44] Some of the novel therapeutic strategies include apoptosis promoter gene transfer, disruption of survival signals by kinase inhibitors, or surface receptor directed apoptosis inducing therapies.[45] Recently it was shown that a cationic nanoparticle (NP) coupled to an integrin alphaVbeta3-targeting ligand can deliver genes selectively to angiogenic blood vessels in tumor-bearing mice. Further the therapeutic efficacy of this approach was tested by generating NP conjugated to a mutant Raf gene, ATPmu-Raf, which blocks endothelial signaling and angiogenesis in response to multiple growth factors. Systemic injection of the NP into mice resulted in apoptosis of the tumor-associated endothelium, ultimately leading to tumor cell apoptosis and sustained regression of established primary and metastatic tumors.[46] Generally, antiangiogenic approaches are less likely to encounter resistance development, but recent studies show there are alternate methods by which tumors may establish channels mimicking vasculatures (vascular mimicry). Vascular mimicry may pose a significant challenge to antiangiogenic therapies such as targeting toxin polypeptides to tumor vasculature. Under these circumstances, complementary approaches such as chemotherapy and radiation can be used to accomplish the complete eradication of tumors.

Acknowledgments

This work was supported in part by a grant from NIH, CA73871, Shirley Ann Sparboe Endowment, Women's Health Fund, and Minnesota Ovarian Cancer Alliance.

References

1. Lyden, D. et al. Impaired recruitment of bone-marrow-derived endothelial and haematopoietic precursor cells blocks tumor angiogenesis and growth (2001) *Nat. Med.* 7, 1194–1201.
2. Aird, W.C., Edelberg, J.M., Weiler-Guettler, H., Simmons, W.W., Smith, T.W., and Rosenberg, R.D. Vascular bed-specific expression of an endothelial cell gene is programmed by the tissue microenvironment (1997) *J. Cell Biol.* 138, 1117–1124.
3. Ferrara, N. VEGF: an update on biological and therapeutic aspects (2000) *Curr. Opin. Biotechnol.* 11, 617–624.
4. Ferrara, N. and Alitalo, K. Clinical applications of angiogenic growth factors and their inhibitors (1999) *Nat. Med.* 5, 1359–1364.
5. De Vries, C., Escobedo, J.A., Ueno, H., Houck, K., Ferrara, N., and Williams, L.T. The fms-like tyrosine kinase, a receptor for vascular endothelial growth factor (1992) *Science*, 255, 989–991.
6. Terman, B.I., Dougher-Vermazen, M., Carrion, M.E., Dimitrov, D., Armellino, D.C., Gospodarowicz, D., and Bohlen, P. Identification of the KDR tyrosine kinase as a receptor for vascular endothelial cell growth factor (1992) *Biochem. Biophys. Res. Commun.* 187, 1579–1586.
7. Shibuya, M., Ito, N., and Claesson-Welsh, L. Structure and function of vascular endothelial growth factor receptor-1 and -2 (1999) *Curr. Top. Microbiol. Immunol.* 237, 59–83.
8. Soker, S., Takashima, S., Miao, H.Q., Neufeld, G., and Klagsbrun, M. Neuropilin-1 is expressed by endothelial and tumor cells as an isoform-specific receptor for vascular endothelial growth factor (1998) *Cell* 92, 735–745.
9. Neufled, G., Cohen, T., Gengrinovitch, S., and Poltorak Z. Vascular endothelial growth factor (VEGF) and its receptors (1999) *FASEB J.* 13, 9–22.
10. Carmeliet, P., Ng, Y.S., Nuyens, D., Theilmeier, G., Brusselmans, K., Cornelissen, I., Ehler, E., Kakkar, V.V., Stalmans, I., Mattot, V. et al. Impaired myocardial angiogenesis and ischemic cardiomyopathy in mice lacking the vascular endothelial growth factor isoforms VEGF164 and VEGF188 (1999) *Nat. Med.* 5, 495–502.
11. Gille, H., Kowalski, J., Yu, L., Chen, H., Pisabarro, M.T, Davis-Smyth, T., and Ferrara, N. A repressor sequence in the juxtamembrane domain of Flt-1 (VEGFR-1) constitutively inhibits vascular endothelial growth factor-dependent phosphatidylinositol 3'-kinase activation and endothelial cell migration (2000) *EMBO J.* 19, 4064–4073.
12. Plate, K.H., Breier, G., Millauer, B., Ullrich, A., and Risau, W. Up-regulation of vascular endothelial growth factor and its cognate receptors in a rat glioma model of tumor angiogenesis (1993) *Cancer Res.* 53, 5822–5827.
13. Baillie, C.T., Winslet, M.C., and Bradley, N.J. Tumor vasculature-a potential therapeutic target (1995) *Br. J. Cancer* 72, 257–267.

14. Denekamp, J. Endothelial cell proliferation as a novel approach to targeting tumor therapy (1982) *Br. J. Cancer* 45, 136–139.
15. LeCouter, J., Kowalski, J., Foster, J., Hass, P., Zhang, Z., Dillard-Telm, L., Frantz, G., Rangell, L., DeGuzman, L., Keller, G.-A., Peale, F., Gurney, A., Hillan, K.J., and Ferrara, N. Identification of an angiogenic mitogen selective for endocrine gland endothelium (2001) *Nature* 412, 877–884.
16. St. Croix, B. et al. Genes expressed in human tumor endothelium (2000) *Science* 289, 1197–1202.
17. Carson-Walter, E.B. et al. Cell surface tumour endothelial markers are conserved in mice and humans (2001) *Cancer Res.* 61, 6649–6655.
18. Huminiecki, L. and Bicknell, R. *In silico* cloning of novel endothelial-specific genes (2000) *Genome Res.* 10, 1796–1806.
19. Huminiecki, L. et al. Magic roundabout is a new member of the roundabout receptor family that is endothelial specific and expressed at sites of active angiogenesis (2002) *Genomics* 79, 547–552.
20. Thorpe, P.E., Chaplin D. J., and Blakey D.C. The First International Conference on Vascular Targeting (2003) *Cancer Res.* 63, 1144–1147.
21. Guo, D., Jia, Q., Song, H.Y., Warren, R.S., and Donner, D.B. Vascular endothelial cell growth factor promotes tyrosine phosphorylation of mediators of signal transduction that contain SH_2 domains (1995) *J. Biol. Chem.* 270, 6729–6733.
22. Darnell, J., Lodish, H., and Baltimore, D. (ed) (1990) *Molecular Cell Biology* (2nd ed.). New York: Scientific American Books, 747–748.
23. Mukherjee, S., Ghosh, R.N., and Maxfield, F.R. Endocytosis (1997) *Physiol. Rev.* 77, 759–803.
24. Ullrich, A and Schlessinger, J. Signal transduction by receptor with tyrosine kinase activity (1990) *Cell* 61, 203–212.
25. Drazin, R., Kandel, J., and Collier, R.J. Structure and activity of diphtheria toxin. II. Attack by trypsin at a specific site within the intact toxin molecule (1971) *J. Biol. Chem.* 246, 1504–1510.
26. Collier, R.J. and Kandel, J. Structure and activity of diphtheria toxin. I. Thiol dependent dissociation of a fraction of toxin into enzymatically active and inactive fragments. (1971) *J. Biol. Chem.* 246, 1496–1503.
27. Wilson, B.A. and Collier, R.J. Diphtheria toxin and *Pseudomonas aeruginosa* exotoxin A: active-site, structure and enzymic mechanism (1992) *Curr. Top. Microbiol. Immunol* 175, 27–41.
28. Ramakrishnan, S., Olson, T.A., Bautch, V.L. and Mohanraj, D. Vascular endothelial growth factor-toxin conjugate specifically inhibits KDR/flk-1-positive endothelial cell proliferation *in vitro* and angiogenesis *in vivo* (1996) *Cancer Res.* 56(6), 1324–1330.
29. Mohanraj, D., Wahlsten, J.L., and Ramakrishnan, S. Expression and radiolabeling of recombinant proteins containing a phosphorylation motif (1996), *Protein. Expr. Purif.* 8, 175–182.
30. Waltenberger, J., Claesson-Welsh, L., Siegbahn, A., Shibuya, M., and Heldin, C.H. (1994) *J. Biol. Chem.* 269, 26988–26995.
31. Arora, N., Masood, R., Zheng, T., Cai, J., Smith, D.L., and Gill, P.S. Vascular endothelial growth factor chimeric toxin is highly active against endothelial cells (1999) *Cancer Res.* 59, 183–188.
32. Olson, T.A., Mohanraj, D., Roy, S., and Ramakrishnan, S. Targeting the tumor vasculature: inhibition of tumor growth by a vascular endothelial growth factor-toxin conjugate (1997) *Int. J. Cancer* 73(6), 865–870.

33. Wild, R., Dhanabal, M., Olson, T.A., and Ramakrishnan, S. Inhibition of angiogenesis and tumor growth by VEGF121-toxin conjugate: differential effect on proliferating endothelial cells (2000) *Br. J. Cancer* 83(8), 1077–1083.

34. Wild, R., Ramakrishnan, S., Sedgewick, J., and Griffioen, A.W. Quantitative assessment of angiogenesis and tumor vessel architecture by computer-assisted digital image analysis: effects of VEGF-toxin conjugate on tumor microvessel density (2000) *Microvasc. Res* 59(3), 368–376.

35. Obrig, T., Louise, C., Lingwood, C., Boyd, B., Barley-Maloney, L., and Daniel, T. Endothelial heterogeneity in Shiga toxin receptors and responses (1993) *J. Biol. Chem.* 268, 15484–15488.

36. Backer, M.V. and Backer, J.M. Functionally active VEGF fusion proteins (2001) *Protein Express. Purif.* 23, 1–7.

37. Backer, M.V. and Backer, J.M. Targeting endothelial cells overexpressing VEGFR-2: selective toxicity of Shiga-like toxin-VEGF fusion proteins (2001) *Bioconjugates* 12(6), 1066–1073.

38. Stirpe, F., Olsnes, S., and Pihl, A. Gelonin, a new inhibitor of protein synthesis, nontoxic to intact cells. Isolation, characterization, and preparation of cytotoxic complexes with concanavalin A (1980) *J. Biol. Chem.* 255, 6947–6695.

39. Veenendaal, L.M., Jin, H., Ran, S., Cheung, L., Navone, N., Marks, J.W., Waltenberger, J., Thorpe, P., and Rosenblum, M.G. *In vitro* and *in vivo* studies of a VEGF121/rGelonin chimeric fusion toxin targeting the neovasculature of solid tumors (2002) *Proc. Natl. Acad. Sci. U.S.A.* 99(12), 7866–7871.

40. Hoffmann, S., Masood, R., Zhang, Y., He, S., Ryan, S.J., Gill, P., and Hinton, R.D. Selective killing of RPE with a Vascular Endothelial Growth Factor chimeric toxin (2000) *IVOS* 41(8), 2389–2393.

41. Schein, C.H. From housekeeper to microsurgeon: the diagnostic and therapeutic potential of ribonucleases (1997) *Nat. Biotechnol.* 15(6), 529–536.

42. Suzuki, M., Saxena, S.K., Boix, E., Prill, R.J., Vasandani, V.M., Ladner, J.E., Sung, C., and Youle, R.J. Engineering receptor-mediated cytotoxicity into human ribonucleases by steric blockade of inhibitor interaction (1999) *Nat. Biotechnol.* 17(3), 265–270.

43. Johnstone, R.W., Ruefli, A.A, and Lowe, S.W. Apoptosis: a link between cancer genetics and chemotherapy (2002) *Cell* 108, 153–164.

44. Fisher, D.E. Apoptosis in cancer therapy: crossing the threshold (1994) *Cell* 78, 539–542.

45. Cory, S. and Adams, J.M., The Bcl2 family: regulators of the cellular life or death switch (2002) *Nat. Rev. Cancer* 2(9), 647–656.

46. Hood, J.D., Bednarski, M., Frausto, R., Guccione, S., Reisfeld, R.A., Xiang, R., and Cheresh, D.A. Tumor regression by targeted gene delivery to the neovasculature (2002) *Science* 296, 2404–2407.

part four

Miscellaneous

chapter ten

---

cBR96-Doxorubicin (SGN-15) for metastatic breast cancer therapy

Lisle Nabell, Clay B. Siegall, and Mansoor N. Saleh

Contents

Introduction

Over the past decade, anticancer therapies have shifted from empiric cytotoxic compounds that exhibit a broad ability to inhibit cell growth to more rationally designed target-based therapies.[1] The goal of this approach is to develop increasingly selective anticancer therapies that avoid the traditional side effects of cytotoxic chemotherapy. Toward this end, antibody-based therapies are becoming increasingly utilized to recognize and direct this type of specific antitumor strategy. Early efforts in this area were often disappointing and hampered by ineffective antibody-mediated cell killing effects, difficulty in achieving tumor specificity, toxicity of the monoclonal

0-4152-6365-4/05/$0.00+$1.50
© 2005 by CRC Press

antibody (mAb), and immunogenicity of the murine mAb. Moreover, the effectiveness of mAbs as mono-therapy in solid tumors, as compared to lymphoid malignancies, has generally been much more limited despite the ability to target specific tumor antigens.[2,3] Reasons for these failures have included the observation that many tumor antigens are not unique to tumors but are shared by normal tissues, leading to increased toxicity in those organs that express the targeted antigen. In addition, the ability of large monoclonal antibodies to navigate into a tumor bed may be inhibited by poor vascularization and increased internal pressure of the tumor relative to the surrounding tissue.[4] Improvements in the design of mAbs have, however, been steady and have included the use of smaller less immunogenic fragments and linkage to a variety of agents.[2,5] SGN-15, a novel immunoconjugate linking the monoclonal antibody BR96, which targets tumors expressing a LewisY (Le^Y)-related antigen and the chemotherapeutic agent doxorubicin represent examples of this approach. Furthermore, the observation that antitumor activity is improved through the combination of conventional chemotherapeutic agents with monoclonal antibodies has suggested a unique strategy to target specific tumors and improve existing therapies.

BR96 monoclonal antibody

BR96 was originally produced as a murine IgG_3, following immunization of a mouse with metastatic human breast carcinoma cells.[6] BR96 is an internalizing antibody which recognizes a carbohydrate structure related to the Le^Y antigen, a protein that is expressed by multiple types of carcinomas, including lung, breast, colon, ovarian, and prostate carcinoma.[6,7] BR96 also binds to normal tissues, most notably the gastrointestinal tract including esophagus, stomach, intestines, and pancreas, with weak binding reported in testis and tonsillar tissue.[6,7] Most of the BR96 antigen is part of a lysosomal membrane glycoprotein 1 (LAMP-1), present on cell-surface domains involved with cellular migration.[8] Following binding of BR96, the antibody is rapidly internalized into endosomes and lysosomes.[7,8] This ability to localize to tumor cells and internalize, where the acidic environment would degrade attached compounds, has made BR96 an attractive antibody for linkage to cytotoxic moieties.

cBR96-doxorubicin (SGN-15)

One strategy to potentiate the tumoricidal activity of mAbs has been conjugation of the antibody to different bioactive substances, including radioactive compounds, toxins, or chemotherapeutic agents. SGN-15 is an example of this approach. SGN-15 is a novel chemoimmunoconjugate composed of the chimeric mAb cBR96, conjugated to doxorubicin using an acid-labile linker (Figure 10.1).[9,10] SGN-15 has a drug/antibody ratio of approximately 8:1 (eight molecules of doxorubicin per antibody molecule). Upon binding to its epitope at the cell surface, SGN-15 is rapidly internalized through endocytosis.

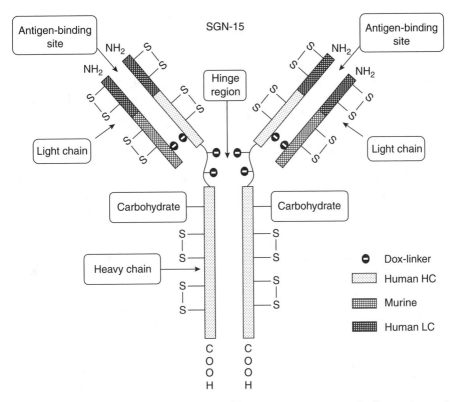

Figure 10.1 As shown schematically above, SGN-15 is a genetically engineered whole chimeric antibody composed of two immunoglobulin heavy chains and two light chains linked by noncovalent interactions, with approximately eight doxorubicin molecules bound via thiol groups to each antibody

Once the compound enters an acidic environment (lysosomes, endosomes), doxorubicin is cleaved by hydrolysis and is able to exert its cytotoxic effects locally. Ideally, this antibody-directed targeting would allow selective release within tumor cells, while sparing normal tissues from typical doxorubicin toxicity. Experiments in animal models have supported this hypothesis, demonstrating significant tumor retention of SGN-15 following intravenous administration.[11]

SGN-15 was initially tested in a number of human tumors grown subcutaneously in athymic nude mice or rats. Administration of SGN-15 either intravenously or intraperitoneally in established tumors resulted in significant partial and complete responses in L2987 lung, RCA colon, and MCF-7 breast tumors.[10,12] In the lung and colon tumor models, SGN-15 effects were dramatic, with complete regressions seen in over 70% of the treated animals.[10] The responses were seen independent of the route of administration or the schedule of infusion (one treatment versus three repeated treatments). This activity was in stark contrast to the results seen with single agent doxorubicin, which did not exhibit any complete responses at optimal doses.[10] These preclinical results suggested that the activity of SGN-15 depended on both the targeting effect of the BR96 antibody and the ability of doxorubicin to exert its effects intracellularly. As mice do not express the Le^Y-related antigen on normal tissues, toxicology studies were carried out in dog models, which demonstrated an acute enteropathy that is likely a result of binding of SGN-15 to Le^Y expressing gastrointestinal epithelial cells.[13] In both the rodent and dog models, there was no evidence of the classic toxicity associated with doxorubicin such as cardiomyopathy.[13]

Phase I studies of SGN-15

Given the activity of SGN-15 seen in preclinical models, phase I clinical trials were undertaken utilizing SGN-15 as a single agent.[14–16] In the first phase I study, a total of 66 patients were treated with an initial schedule of SGN-15 administered over 2 h every 3 weeks.[14] Although the study included a variety of tumor types, the majority of the patients enrolled on this trial had metastatic breast or colorectal carcinoma, with high expression of the BR96 antigen.[14] This study defined the gastrointestinal side effects of SGN-15 as dose limiting, with the most common side effects consisting of nausea, vomiting, and gastritis.[14,16] At higher doses, hematemesis occurred with evidence of exudative gastritis observed on endoscopy.[14] Due to the occurrence of significant gastrointestinal toxicity at higher doses, a variety of approaches were employed to limit the adverse events. These strategies included aggressive premedication consisting of corticosteroids, a proton pump inhibitor, and a 5-hydroxy-tryptamine-3 antagonist, and institution of a 24-h infusion schedule. These changes resulted in modest improvement in the observed side effects.[14] Additional side effects seen included fever, asthenia, and temporary elevation of serum lipase and amylase but without symptoms of clinical pancreatitis.[14,16] Of interest, there were no events that reflected the more traditional toxicities of doxorubicin such as neutropenia, alopecia, or cardiomyopathy. The gastrointestinal toxicity encountered in this phase I study was likely due to binding of the immunoconjugate to normal tissue expressing the Le^Y antigen, as opposed to local effects of doxorubicin. Supporting evidence for this observation came from Saleh et al. who reported that administration of unconjugated chimeric BR96 antibody at a dose of 550 mg/m^2 to one patient resulted in bloody emesis and bloody diarrhea, similar in appearance to the effect of higher doses of SGN-15.[14] The phase II dose identified for every 21-day administration of SGN-15 from this trial was 700 mg/m^2 (21 mg/m^2 of doxorubicin). In five patients with easily assessable tumors, biopsies were performed. In all cases, immunohistochemistry localized BR96 to tumor tissue with evidence of intranuclear deposition of doxorubicin, confirming that molecular targeting was achieved.[14]

A second phase I trial of SGN-15 was undertaken employing a weekly schedule.[15] Patients with Le^Y-expressing tumors were enrolled to receive SGN-15 on an escalating schedule weekly for the first 3 weeks of a 5-week cycle. A total of 42 patients were entered on this trial. Dose limiting toxicity mirrored that of the first phase I study with gastrointestinal toxicity including vomiting, exudative gastritis and hematemesis reported. Based on these results, the recommended weekly dose for phase II studies was 250 mg/m^2 (7 mg/m^2 of doxorubicin).[15]

Pharmacokinetic studies of SGN-15

One of the concerns of using an immunoconjugate such as SGN-15 has been the possibility of unwanted hydrolysis, resulting in circulating free doxorubicin. Evidence that this does not occur has included the observation that the typical toxicities of doxorubicin are not reported in the initial phase I studies. Further support that little serum release of doxorubicin occurred comes from studies by Saleh et al. who quantitated total BR96 and doxorubicin by ELISA and HPLC following administration of SGN-15 on a 21-day schedule.[14] Total amount of free doxorubicin found by HPLC in 66 patients treated with SGN-15 was reported to be minute.[14] The possibility of developing a hypersensitivity response to repeated antibody exposure has also been a concern, although a consistent finding through the phase I/II trials has been a lack of significant hypersensitivity reactions or evidence of a significant immunologic response, despite some patients having received multiple administrations of SGN-15.[14,17,18]

Phase II studies of SGN-15 in breast cancer

Of the patients in the initial phase I studies, two experienced a partial response: one with breast and another with gastric cancer. These results prompted continued interest in the activity of SGN-15, and a subsequent phase II trial was undertaken in patients with metastatic breast cancer.[17] Tolcher et al. reported the results of a randomized, multicenter, phase II trial comparing single agent doxorubicin at 60 mg/m^2 every 3 weeks versus SGN-15 at 700 mg/m^2 given every 3 weeks in patients with metastatic breast cancer. Patients were allowed to receive up to a maximum of six courses of therapy with crossover to the alternative regimen if they experienced progressive disease or had persistently stable disease. A total of 25 patients were enrolled on the study. This group was not heavily pretreated, with only one prior chemotherapy administration in the metastatic setting allowed.[17] Based on toxicity results from the phase I trial, SGN-15 was administered as a continuous intravenous infusion over 24 h while single agent doxorubicin was given as an intravenous push over 15 min. Despite the protracted infusion schedule, patients treated with SGN-15 experienced significant gastrointestinal toxicity including nausea, vomiting, and gastritis, with two patients developing hematemesis. The response to SGN-15 was judged to be limited with only one patient having a partial response compared to four patients who responded to single agent doxorubicin.[17] Six patients experienced disease stabilization following treatment with SGN-15, compared to five patients receiving doxorubicin. Of interest however, were the results seen in patients who crossed over to receive SGN-15 after exhibiting stable or progressive disease on single agent doxorubicin. Two of four patients treated in this fashion demonstrated a partial response to SGN-15 after crossover, suggesting a lack of cross-resistance between SGN-15 and doxorubicin.[17]

In vitro *activity of SGN-15 in combination with chemotherapy*

Targeted immunotherapy using mAbs for solid tumors has had limited success when employed as a single agent. The factors underlying this lack of efficacy likely represent myriad findings, but have been thought to include decreased functional vasculature, lack of antitumor specificity, and elevated interstitial pressure.[4] Attempts to improve the clinical efficacy of mAbs have included the strategy of combining with more conventional cytotoxic agents. The concept of enhancement of response through the use of combination therapy has been successful in several areas including the concomitant use of radiation therapy with chemotherapy in the treatment of lung cancer and head and neck cancer, and in the application of mAb with traditional chemotherapy agents. In the case of breast cancer, where the mAb Herceptin is available, single agent activity is limited to around 14%.[19] However, when Herceptin is combined with traditional chemotherapy agents such as paclitaxel or carboplatin, reponses are much higher, ranging from 50% to 70%.[19,20] This enhancement of response may occur even with the use of suboptimal doses of chemotherapy due to the ability of one agent to potentiate the effects of the second. For example, small doses of taxanes that lack significant systemic effect can be used as radiosensitizors, as they help block tumor cells in G2/M, a portion of the cell cycle that is exquisitely sensitive to the application of radiation.[21] While SGN-15 demonstrated activity as a single agent, gastrointestinal toxicity at the doses necessary to produce the antitumor effect in humans has precluded its usage as a single agent. Based on evidence in other models of synergy with combinations of chemotherapy and targeted immunotherapy (i.e., Rituxan or Herceptin in combination with chemotherapy), preclinical studies were conducted to evaluate this approach with SGN-15.

Using several different human carcinoma cell lines including breast, lung, and colon, investigators confirmed significant synergistic antitumor effect with the application of SGN-15 prior to exposure to a taxane (paclitaxel or docetaxel).[22,23] This suggestion of benefit was confirmed in subsequent xenograft studies of human lung, colon, and breast tumors where combination therapy with paclitaxel and SGN-15 resulted in significant antitumor activity that was superior to either agent used alone.[22] Moreover, this enhanced response was seen both in taxane-sensitive and -insensitive tumor models. This improved response was clearly due to the intratumoral presence of doxorubicin as unconjugated cBR96 or free doxorubicin administered with paclitaxel did not provide the same benefit.[22]

The success of combination therapy likely depends on the mechanism by which these agents influence cell cycle kinetics. Both SGN-15 and the taxanes can affect cell cycle progression and could lead to detrimental as well as synergistic effects. The taxanes (including paclitaxel and docetaxel) are known to affect polymerized tubulin, promoting microtubule formation and inhibiting disassembly.[24] These properties result in the disruption of microtubule function, and could impair the endocytotic pathway necessary for SGN-15 internalization. Initial studies of human cancer cell lines suggested that pretreatment with SGN-15 resulted in accumulation of cells in G2, with increased sensitivity to paclitaxel. The use of an alternative taxane, docetaxel, produced similar results. Confirmation of the benefit of sequential SGN-15 followed by taxane exposure was seen in human cancer xenograft models, where the staggered combination approach produced significant antitumor effects at doses of SGN-15 that would have been considered ineffective.[23] The need to perform sequential administration of the agents was emphasized by evidence of antagonistic effects if the drugs were given in reverse order or simultaneously.[23]

Phase I clinical trials of SGN-15 with chemotherapy

Based on preclinical data, which suggested lower doses of SGN-15 could be effective when given in combination with a taxane, a phase I/II trial using SGN-15 and docetaxel was undertaken.[25] In the phase I portion of the trial, a total of 16 patients with metastatic breast or colorectal cancer were enrolled. As part of the eligibility criteria, the tumors were tested to ensure that they expressed the BR96-related antigen. Patients were treated with escalating doses of SGN-15 followed by a taxane, docetaxel, on a weekly basis for 6 weeks followed by reassessment. The initial starting dose of SGN-15 was $100\,mg/m^2$ while patients received a fixed weekly dose of docetaxel at $30\,mg/m^2$. Dose escalation occurred up to $200\,mg/m^2$ of SGN-15. Gastrointestinal toxicity was the most prominent toxicity observed, with one case of clinical pancreatitis and one case of hemorrhagic gastritis. Other toxicities observed with this combination therapy included nausea, vomiting, neutropenia, and hyperbilirubinemia. There were no cases of protracted nausea or vomiting following treatment and no evidence of cardiac toxicity was seen. One patient achieved an objective tumor response, which was continued through a total of six courses of therapy, the maximum allowed on the study. A second patient initially responded to therapy but developed progressive disease at the end of two courses. A third patient who discontinued therapy due to gastrointestinal symptoms was noted to have stable disease. Based on the observation that the $150\,mg/m^2$ dose was well tolerated but gastrointestinal toxicity was encountered at the $200\,mg/m^2$ dose, a slightly lower dose of $175\,mg/m^2$ of SGN-15 was chosen for the subsequent phase II study.[25]

Phase II study of SGN-15 with docetaxel in metastatic breast cancer

The combination of SGN-15 with docetaxel has been evaluated in a phase II multicenter study in patients with metastatic breast cancer, with results available only in abstract form at this date.[26] Following the format from the phase I study, patients with metastatic Le^Y-expressing metastatic breast cancer received SGN-15 (175 mg/m^2) and docetaxel (30 mg/m^2) for 6 weekly infusions followed by a 2-week rest period (8-week course). A total of 30 patients enrolled on the trial with 27 evaluable for response. Five of the evaluable patients experienced an objective response (1 partial response and 4 with minimal response). Nine patients were found to have stable disease after their first course of therapy. The majority of the grade 3 toxicities were gastrointestinal in nature.[26] The results of this study suggest that SGN-15, in combination with docetaxel, has activity in the treatment of breast cancer but that, even when combined with traditional chemotherapy, the gastrointestinal side effects limit the clinical application of this drug.

Summary

SGN-15 represents a unique mechanism of target-based immunotherapy, using a monoclonal antibody to deliver intracellular doxorubicin. Although initial animal models demonstrated striking activity, the observed gastrointestinal activity has precluded the ability to deliver large doses. The clinical activity of SGN-15, as a single agent or in combination with docetaxel, when tested in patients with colon or breast cancer has been limited. Of interest is the observation that the gastrointestinal toxicity seen in breast cancer trials was not observed in a recent study of SGN-15 in lung cancer patients.[27] Work in the area of targeted immunotherapy with SGN-15 is being undertaken with other tumor types, and merits further investigation.

References

1. Balis FM. Evolution of anticancer drug discovery and the role of cell-based screening. *J Natl Cancer Inst* (2002) 94:78–9.
2. Hellstrom I, Trail P, Siegall C, Firestone R, and Hellstrom KE. Immunoconjugates and immunotoxins for therapy of solid tumors. *Cancer Chemother Pharmacol* (1996) 38(Suppl):S35–6.
3. Treon SP, Shima Y, Grossbard ML, et al. Treatment of multiple myeloma by antibody mediated immunotherapy and induction of myeloma selective antigens. *Ann Oncol* (2000) 11(Suppl 1):107–11.
4. Jain RK. Barriers to drug delivery in solid tumors. *Sci Am* (1996) 271:58–65.
5. Yarnold S and Fell HP. Chimerization of antitumor antibodies via homologous recombination conversion vectors. *Cancer Res* (1994) 54:506–12.
6. Hellstrom I, et al. Highly tumor-reactive, internalizing mouse monoclonal antibodies to Le, related cell surface antigens. *Cancer Res* (1990) 50:2183–90.

7. Garrigues J, et al. LeY specific antibody with potent anti-tumor activity is internalized and degraded in lysosomes. *Am J Pathol* (1993) 142:607–22.

8. Garrigues J, Anderson J, Hellstrom KE, and Hellstrom I. Anti-tumor antibody BR96 blocks cell migration and binds to a lysosomal membrane glycoprotein on cell surface microspikes and ruffled membranes. *J Cell Biol* (1994) 125(1): 129–42.

9. Willner D, Trail PA, Hofstead SJ, et al. (6-Maleimidocaproyl) hydrazone of doxorubicin—a new derivative for the preparation of immunoconjugates of doxorubicin. *Bioconjug Chem* (1994) 4:521–7.

10. Trail PA, Willner D, Lasch SJ, et al. Cure of xenografted human carcinomas by BR96-doxorubicin immunoconjugates. *Science* (1993) 261:212–15.

11. Mosure KW, Henderson AJ, Klunk LJ, and Knipe JO. Disposition of conjugate-bound and free doxorubicin in tumor-bearing mice following administration of a BR96-doxorubicin immunoconjugate (BMS 182248). *Cancer Chemother Pharmacol* (1997) 40:251–25.

12. Sjogren HO, Isaksson M, Willner D, et al. Antitumor activity of carcinoma-reactive BR96-doxorubicin conjugate against human carcinomas in athymic mice and rats and syngeneic rat carcinomas in immunocompetent rats. *Cancer Res* (1997) 57:4530–6.

13. Saleh M, LoBuglio A, Sugarman S, et al. Gastrointestinal effects of chimeric BR96-doxorubicin (DOX) conjugate. *Proc Am Assoc Cancer Res* (1995) 36:A2876 (abst).

14. Saleh MN, Sugarman S, Murray J, et al. Phase I trial of the anti-Lewis Y drug immunoconjugate BR96-doxorubicin in patients with Lewis Y-expressing epithelial tumors. *JCO* (2000) 18(11):2282–92.

15. Giantonio TA, Gilewski TA, Bookman MA, et al. A phase I study of weekly BR96-doxorubicin (BR96-Dox) in patients (pts) with advanced carcinoma expressing the Lewis[Y] (Le[Y]) antigen. *Proc ASCO* (1996) 15:433 (abst 1380).

16. Sugarman S, Murray JL, Saleh M, et al. A phase I study of BR96-doxorubicin (cBR96-Dox) in patients with advanced carcinoma expressing the Lewis (Y) antigen. *Proc Am Soc Clin Oncol* (1995) 14:A1532 (abst).

17. Tolcher AW, Sugarman S, Gelmon KA, et al. Randomized phase II study of BR96-doxorubicin conjugate in patients with metastatic breast cancer. *J Clin Oncol* (2000) 18:2282–92.

18. Nabell L, Saleh M, Marshall J, Hart L, et al. A phase II study of SGN-15 (cBR96-doxorubicin immunoconjugate) combined with docetaxel for the treatment of metastatic breast and colorectal carcinoma. *Proc ASCO* (2002) 21:55a.

19. Slamon D, Leyland-Jones B, et al. Addition of herceptin (humanized anti-her2 antibody) to first line chemotherapy for her2 overexpressing metastatic breast cancer markedly increases anticancer activity: a randomized multinational controlled phase III trial. *New Engl J Med* (2001) 344:783–92.

20. Robert NJ, Leyland-Jones B, Asmar L, et al. Phase III comparative study of trastuzumab and paclitaxel with and without carboplatin in patients with Her-2/neu positive advanced breast cancer. *Breast Cancer Res Treat* (2002) 76 (Suppl):35a.

21. Rowinsky EK, and Donehower RC. Paclitaxel (taxol) [published erratum appears in *N Engl J Med* 1995 July 6:333:75]. *N Engl J Med* (1995) 332:1004–14.

22. Trail PA, Willner D, Bianchi AB, et al. Enhanced antitumor activity of paclitaxel in combination with the anticarcinoma immunoconjugate BR96-Doxorubicin. *Clin Cancer Res* (1999) 5:3632–8.

23. Wahl AF, Donaldson KL, Mizan BJ, Trail PA, and Siegall CB. Selective tumor sensitization to taxanes with the mAb-drug conjugate cBR96-doxorubicin. *Int J Cancer* (2001) 93:590–600.

24. Hanauske AR, Degen D, Hilsenbeck SG, et al. Effects of taxotere and taxol on *in vitro* colony formation of freshly explanted human tumor cells. *Anticancer Drugs* (1992) 3:121.

25. Nabell L, Sing AP, Siegall AP, et al. A phase 1 study of combined modality therapy using SGN-15 in combination with taxotere for the treatment of metastatic breast and colorectal carcinoma. *Proc ASCO* (2001) 20:2251a.

26. Hart L, Nabell L, Saleh M, et al. Phase II study of SGN-15 (cBR96-doxorubicin immunoconjugate) combined with docetaxel for the treatment of metastatic breast carcinoma. *Proc ASCO* (2003) 22:696a.

27. Ross H, Rudin CM, Hart L, Figlin RA, et al. Phase II study of SGN-15 (cBR96-doxorubicin immunoconjugate) combined with docetaxel in patients with advanced stage or metastatic non-small cell lung cancer (NSCLC). *Proc ASCO* (2003) 22:690a.

chapter eleven

Anti-CD33 calicheamicin immunoconjugate therapy of acute myeloid leukemia

Charalambos Andreadis, Selina M. Luger, and Edward A. Stadtmauer

Contents

0-4152-6365-4/05/$0.00+$1.50
© 2005 by CRC Press

Introduction

Acute myeloid leukemia

Acute myeloid leukemia (AML) is the most frequent form of leukemia in adults and is projected to be the seventh leading cause of cancer death in the U.S. in 2003 (American Cancer Society 2003: 10). The age-adjusted incidence in the U.S. population was 3.7 per 100,000 per year, between 1996 and 2000 (Ries et al. 2003). Incidence increases with age, from 1.8 per 100,000 per year in those less than 65 years of age to 17.2 per 100,000 per year in those 65 and over (Ries et al. 2003). Incidence is higher in men (4.6 per 100,000 per year) than in women (3.1 per 100,000 per year) across all age groups (Ries et al. 2003).

Among patients younger than 60 years with de novo AML, complete remission (CR) can be routinely induced in 60% to 80% of those treated with a combination of an anthracycline, such as idarubicin, daunorubicin, or mitoxantrone and infusional cytarabine (ara-C) (Anonymous 1998; Arlin et al. 1990; Berman et al. 1991; Cassileth et al. 1998; Mayer et al. 1994; Rees et al. 1986; Wiernik et al. 1992). Among complete responders, 40% to 60% can be cured with current postremission therapy in this age group (Cassileth et al. 1998; Löwenberg et al. 1999; Mayer et al. 1994). Young patients who relapse generally have a response to salvage therapy ranging between 30% and 70%, with higher doses of ara-C, or second-line anthracycline-based regimens (Archimbaud et al. 1995; Davis et al. 1993; Keating et al. 1989; Kern et al. 1998; Vogler et al. 1994). Cure rates after a second remission range between 10% and 20% with conventional therapy (Archimbaud et al. 1995; Geller et al. 1989; Keating et al. 1989; Webb 1999) and 20% to 30% after stem cell transplantation (SCT) (Clift et al. 1987; Gale et al. 1996; Geller et al. 1989).

Adults aged 60 years and older have inferior outcomes, affected by the higher incidence of unfavorable cytogenetic abnormalities (Baudard et al. 1994; Hiddemann et al. 1999; Leith et al. 1997; Rowe 2000; Rowley et al. 1982), the presence of preexisting myelodysplasia (Baudard et al. 1994; Hiddemann et al. 1999; Rowe 2000) and the increased prevalence of the multidrug resistance phenotype (MDR) (Hiddemann et al. 1999; Leith et al. 1997). Moreover, older adults tend to have more medical comorbidities and tolerate intensive chemotherapy less well than their younger counterparts. In this age group, complete remission is only achieved in 40% to 60% of those who are candidates for induction chemotherapy (Anonymous 1998; Baudard et al. 1994; Hiddemann et al. 1999; Leith et al. 1997; Löwenberg et al. 1998; Rowe 2000). Among complete responders, approximately 20% survive free of leukemia for at least 2 years (Leith et al. 1997; Löwenberg et al. 1998; Löwenberg et al. 1999). Those who relapse find further treatment options to be limited, often opting for palliation of symptoms rather than cure.

CD33 antigen

CD33 was first described as an antigenic determinant on human myeloid progenitor cells and their progeny, recognized by the murine monoclonal antibodies L4F3 (Andrews et al. 1983) and MY9 (Griffin et al. 1984). It is a 67 K_d transmembrane glycoprotein that is a member of the sialic acid-binding immunoglobulin-related lectin family (Freeman et al. 1995). During normal hematopoiesis, it is expressed on myeloid, erythroid, and megakaryocytic committed progenitor cells (Andrews et al. 1983; Andrews et al. 1986; Griffin et al. 1984; Pierelli et al. 1993; Tanimoto et al. 1989) but not on uncommitted pluripotent hematopoietic stem cells (Andrews et al. 1983; Andrews et al. 1989; Bernstein et al. 1987; Pierelli et al. 1993; Wagner et al. 1995). It maintains low-level expression on peripheral granulocytes and tissue macrophages (Pierelli et al. 1993; Tanimoto et al. 1989) but not on erythrocytes, lymphocytes, platelets, or non-hematopoietic tissues (Griffin et al. 1984; Pierelli et al. 1993). The CD33 antigen is expressed on approximately 90% of AML myeloblasts, including leukemic clonogenic precursor cells believed to be derived from normal hematopoietic progenitor cells (Dinndorf et al. 1986; Griffin et al. 1984; Legrand et al. 2000; Sabbath et al. 1985; Scheinberg et al. 1989). Overall, CD33 intensity in total CD33+ cells is significantly higher in the bone marrow than in peripheral blood (Jilani et al. 2002). Additionally, patients with AML younger than 60 years of age had significantly higher intensity of CD33 expression on CD33+ cells than patients 60 years or older (Jilani et al. 2002).

Antibodies in the therapy of AML

The ready availability of monoclonal antibodies directed against hematopoietic differentiation antigens and the expression of these antigens on leukemic cells and their precursors, inaugurated early studies of leukemia immunotherapy in murine models (Bernstein et al. 1980; Kirch et al. 1981). Subsequent pilot trials conducted in patients with leukemia have demonstrated the ability of these monoclonal antibodies to bind target cells and induce variable responses (Ball et al. 1983; Dillman et al. 1984; Foon et al. 1984; Nadler et al. 1980; Ritz et al. 1981). A challenge in treating AML with monoclonal antibodies has been the necessity for selection of an antigenic determinant found on myeloblasts and their malignant precursors but not on the ultimate hematopoietic progenitors or mature circulating cells (Scheinberg et al. 1991).

Based on its expression profile, CD33 appeared to be a desirable target, and an early experiment demonstrated that *in vitro* treatment of cells from certain patients with AML with L4F3 can reconstitute normal early hematopoiesis (Bernstein et al. 1987). Antibodies directed against CD33 have since been utilized to purge bone marrow grafts *in vitro* and *in vivo* in the treatment of AML (Appelbaum et al. 1992; Lemoli et al. 1991; Robertson et al. 1992). However, antigenic modulation of CD33 through antigen–antibody complex internalization (Scheinberg et al. 1989; Tanimoto et al. 1989; Ulyanova et al. 1999) has limited the direct cytotoxicity of anti-CD33 antibodies and has instead established them as effective vehicles for radiation and toxin delivery to target cells (Caron et al. 1998; Jurcic et al. 2000; Jurcic et al. 2002; Pagliaro et al. 1998; Roy et al. 1991; Schwartz et al. 1993).

The development of gemtuzumab ozogamicin

Calicheamicins, produced by the fungus *Micromonospora echinospora* ssp. *calichensis*, were discovered during a search for new fermentation-derived antitumor antibiotics. They show extraordinary potency against murine tumors and are 1000- to 4000-fold more potent than adriamycin (Lee et al. 1987a; Lee et al. 1987b). They represent the novel class of enediyne antitumor antibiotics, composed of naturally occurring and synthetic compounds with a common structural motif (Nicolaou et al. 1993). Calicheamicin γ_1^I, the prototype compound in this class, is a highly active DNA cleaving agent that introduces sequence-selective double-stranded cuts (Ikemoto et al. 1995; Nicolaou et al. 1993; Zein et al. 1988; Zein et al. 1989). A hydrazide derivative of calicheamicin, NAc-γ-calicheamicin DMH with potent antitumor activity *in vitro* was linked to a humanized version (IgG4) of the anti-CD33 antibody, anti-hP67.6, to generate CMA-676 or gemtuzumab ozogamicin (GO, Mylotarg™) (Hamann et al. 2002a; Hamann et al. 2002b; Hinman et al. 1998) (Figure 11.1). The amide linker is hydrolyzed upon internalization of the CD33 antigen–antibody complex to the acidic micro-environment of the endosome, to release the activated calicheamicin derivative (Hamann et al. 2002a; Hamann et al. 2002b). GO has shown remarkable selective activity in leukemic cell lines (Amico et al. 2003; Hamann et al. 2002b; Linenberger et al. 2001; Matsui et al. 2002; Naito et al. 2000; Walter et al. 2003) with evidence of effect attenuation by MDR1 (Linenberger et al. 2001; Matsui et al. 2002; Naito et al. 2000; Walter et al. 2003) as well as DNA-damage response elements Chk1/Chk2 and Caspase 3 (Amico et al. 2003).

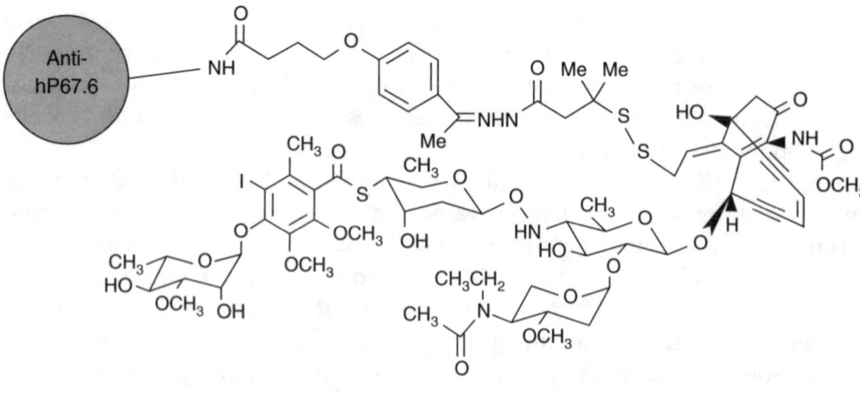

Figure 11.1 NAc-γ-calicheamicin is covalently linked to the humanized monoclonal antibody anti-hP67.6 via a hydrolyzable amide linker. Approximately 50% of the antibody is conjugated with an average of two to three molecules of calicheamicin per antibody molecule

Clinical utility

Pharmacology

At the dose of 9 mg/m^2 used in phase II studies, GO exhibits the highest binding affinity for myeloid blast cells and monocytes, intermediate affinity for mature granulocytes and minimal affinity for lymphocytes (van der Velden et al. 2001). Near complete saturation of CD33 antigenic sites is observed 30 min to 4 h after a single GO infusion resulting in rapid antigen–antibody complex internalization, and ultimately, the induction of apoptosis in CD33+ myeloid cells (Sievers et al. 1999; van der Velden et al. 2001). Moreover, continuous renewed expression of CD33 antigen is observed on the surface of myeloid cells, significantly increasing the amount of internalized antibody (van der Velden et al. 2001). Interestingly, the mean maximal GO binding to myeloid blasts was found to be lower at infusion cycle 2 than at cycle 1, suggesting that myeloblasts with high-level CD33 expression may undergo accelerated clearance from the circulation (van der Velden et al. 2001). The plasma half-life of the vector antibody, anti-hP67.6, which appears to be representative of the molecule as a whole, is 38 to 72 h after a single dose of GO at 9 mg/m^2 (Korth-Bradley et al. 2001; Sievers et al. 1999). Of note, pharmacokinetic parameters are similar across groups defined by sex and a cutoff age of 60 years (Korth-Bradley et al. 2001).

Preapproval trials

Efficacy

Forty patients with relapsed or refractory AML were treated during a phase I protocol (Sievers et al. 1999) with up to three infusions of GO ranging in dose from 0.25 to 9 mg/m^2 at an interval of at least 14 days. Twenty-two (71%) patients with peripheral blood myeloblasts experienced a reduction in their peripheral blast counts, while eight patients (20%) achieved a bone marrow response (less than 5% leukemic blasts on morphologic examination). Three of them went on to achieve a complete response by NCI criteria (CR), including one patient treated at the 1 mg/m^2 dose level. The 9 mg/m^2 dose level was selected for the phase II studies based on the lack of dose-limiting toxicity at that dose. Additionally, GO infusions were generally limited to two (instead of three) to prevent prolonged neutropenia seen at that dose intensity.

In three international phase II studies (Sievers et al. 2001), 142 patients with primary, CD33+ AML were treated in first relapse, after maintaining a complete response for 3 months or longer, five of them after transplantation (Table 11.1). Based on reports of fatal pulmonary toxicity and tumor lysis syndrome in the setting of elevated blast counts, a white blood cell count of less than 30,000/μl was required for entry into the phase II studies. Treatment with hydroxyurea or leukapheresis was acceptable to reduce white cell

Table 11.1 Experience with gemtuzumab ozogamicin, alone or in combination, as induction therapy in acute myeloid leukemia and high-grade myelodysplastic syndrome

Author	Disease	Setting	n	Therapy	Younger OR (CR) (%)	Older OR (CR) (%)	Grade 3/4 liver toxicity (VOD) (%)
Sievers et al.[a] 2001	AML	Relapsed or refractory	142	GO (9 mg/m² d 1,15)	34	26	38 (3)
Larson et al.[b] 2002b	AML	Relapsed or refractory	101	GO (9 mg/m² d 1,15)		28 (13)	39 (5)
Amadori et al. 2002	AML	De novo	57	I: GO (9 mg/m² d 1,15)		35 (22)	14 (2)
				II: Mitoxantrone (7 mg/m² d 1,3,5) Ara-C (100 mg/m² d 1–7) Etoposide (100mg/m² d 1–3)		54 (35)	— (4)
Estey et al. 2002	AML MDS	De novo	51	GO (9 mg/m² d 1/8–15) ±IL-11 (15 µg/kg d 3–28)		22 (22)	— (16)
Cohen et al. 2002	AML	Relapsed post-SCT	8	GO (9 mg/m² d 1,15)	25[c] (0)		25 (12.5)
Alvarado et al. 2003	AML	Relapsed or refractory	14	GO (6 mg/m² d 1,15) Idarubicin (12 mg/m² d 2–4) Ara-C (1.5 g/m² d 2–5)	43 21		19 14
Tsimberidou et al. 2003	AML MDS	De novo	59	GO (6 mg/m² d 1) Fludarabine (15 mg/m² d 2–6) Cyclosporine (22 mg/kg d1/16 mg/kg d2) Ara-C (1 g/m² d 2–6)	47 (46)		38 (7)
Cortes et al. 2002	AML MDS	Relapsed or refractory	17	GO (9 mg/m² d 1) Topotecan (1.25 mg/m² d 1–5) Ara-C (1 g/m² d 1–5)	12 (12)		41 (6)
De Angelo et al. 2003	AML	De novo	43	GO (6 mg/m² d 4) Daunorubicin (45 mg/m² d 1–3)	84 84		30 (10[d])

Study	Diagnosis	Status	N	Regimen				
Baccarani et al. 2002	AML	De novo	7	GO (6 mg/m² d 1/4) Ara-C (100 mg/m² d 1–7)				
Kell et al. 2002	AML	De novo	55	GO (3 mg/m² d 1) *with* Daunorubicin (50 mg/m² d 1,3,5) Ara-C (100 mg/m² d 1–10) Thioguanine (100 mg/m² d 1–10) *or* Daunorubicin and Ara-C as above *or* Fludarabine (30 mg/m² d 1–5) Ara-C (2 g/m² d 1–5) Idarubicin (10 mg/m² d 1–3) G-CSF (5 µg/kg d 6 on)	85 (85)	57 (57)	8 (0)	42 (4)
Litzow et al. 2003	AML	Relapsed of refractory	26[e]	GO (6 mg/m² d 5) Ara-C (1 g/m² d 1–5)	12.5 (12.5)			

AML, acute myeloid leukemia; CR, complete response; GO, gemtuzumab ozogamicin; MDS, myelodysplastic syndrome; OR, overall response; SCT, stem cell transplantation; VOD, veno-occlusive disease of the liver.

[a] Later updated by Larson et al. 2002a.
[b] Subset in previously reported trial by Sievers et al. 2001.
[c] One unconfirmed CRp.
[d] VOD after subsequent SCT.
[e] Enrolled in GO arm of the study.

counts below the eligibility threshold. The median age of the patients was 61, with a median duration of prior CR of 11.1 months. Most of them (94%) had received postremission therapy, with 70% having received high-dose ara-C containing regimens. Only 5% of these patients were in the favorable-cytogenetic risk category at the time of relapse. Overall, 46% achieved a bone marrow response after a single dose of GO. Across all studies, 30% of patients achieved an overall response (OR), with 16% CR and 13% CRp (complete response with incomplete platelet recovery). The latter patients met the criteria of a CR, except that of platelet recovery, although platelet transfusion independence was required for at least a week. The median time to an absolute neutrophil count (ANC) greater than 1500/µl was 41 days and the median time to response was 60 days. Median relapse-free survival was 6.8 months (7.2 months for those in CR and 4.4 months for those in CRp). Overall survival of the group at 1 year was 31%. Of note, achievement of overall response was similar across groups defined by age, cytogenetics, or duration of first remission. Additionally, more than 38% of participants received a dose of GO as outpatients, with 24% receiving both doses this way.

The final report on 277 patients enrolled in the phase II trials (Larson et al. 2002a) confirmed the above numbers, with an overall response of 26% (13% CR and 13% CRp). Again, response was similar across groups defined by age and cytogenetics (Stadtmauer et al. 2002). However, relapse-free survival was significantly better at 1 year for patients younger than 60 years (36%) compared to that of older patients (7%) (Wyeth, unpublished data).

Twenty-five patients with relapsed/refractory AML enrolled in the phase II studies went on to receive SCT after attaining a CR2 or CRp2 with GO (Sievers et al. 2002). The median relapse-free survival in this group was 10.7 months, compared to 5.3 months for the group of responders as a whole. The median overall survival after SCT was 15.6 months in this group. Based on these results, a phase III Eastern Cooperative Oncology Group (ECOG) trial is underway to evaluate the utility of GO prior to autologous stem cell transplantation in first remission AML.

Older patients

Overall, 101 patients with relapsed/refractory AML, aged 60 years and older (23% of whom were 70 years and older) participated in the phase II trials (Table 11.1). Of note, none of these patients had secondary AML or AML in the setting of myelodysplastic syndrome (MDS), both of which are more prevalent in the elderly (Larson et al. 2002b). Ninety-three percent had received postremission therapy after their first remission, 44% with intermediate- or high-dose ara-C. Twenty per cent of patients received one dose, 76% two and 4% three doses of GO. Overall response rate was 28% (13% CR and 15% CRp), with a median time to remission of 65 days for those in CR and 75 days for those in CRp. Median overall survival among responders was 14.5 months for patients in CR and 11.8 months for patients in CRp, with an overall survival across groups at 5.4 months. Five patients received a second course of therapy with two responses and a patient received a third course of therapy with a subsequent response. Induction-related toxicities were similar to younger patients, as well as elderly patients in the prior studies. Based on these results, GO was approved by the FDA for use in elderly patients with CD33+ AML in first relapse (Bross et al. 2001).

Toxicity

Postinfusion syndrome was the most frequent non-hematologic toxicity observed after GO infusion, with 7% of the patients experiencing grade 3 or 4 fever, 11% grade 3 or 4 chills, and 4% grade 3 or 4 hypotension in the phase II trials (Sievers et al. 2001). This occurred despite premedication with acetaminophen and diphenhydramine but was generally attenuated with subsequent GO administrations.

After receipt of two infusions in the phase II studies, grade 3 or 4 induction related toxicities included the predictable neutropenia (97%), thrombocytopenia (99%) as well as bleeding (15%), mucositis (4%), and infections (28%) (Sievers et al. 2001). There were no reported cardiac toxicities or alopecia. Thirteen per cent of patients died during induction therapy. The causes of death included progression of disease (42%), CNS hemorrhage (26%), multiorgan failure (21%), and sepsis (11%).

No humoral immune responses to anti-hP67.6 antibody were detected in the phase I and II studies (Sievers et al. 1999; Sievers et al. 2001). A humoral response to the calicheamicin–linker complex was documented in one patient after receiving a third dose and in a second patient upon re-treatment with GO in the phase I study (Sievers et al. 1999).

Veno-occlusive disease

Hepatic veno-occlusive disease (VOD) is a well-recognized clinical syndrome occurring most frequently in the setting of high-dose cytoreductive therapy and hematopoietic stem cell transplantation (Jones et al. 1987; McDonald et al. 1993) but occasionally described after conventional chemotherapy (Barclay et al. 1994; Gill et al. 1982; Ortega et al. 1997). The outcome of severe disease is poor, with death from progressive multiorgan failure being the usual consequence (Jones et al. 1987; McDonald et al. 1993). Treatment with GO has been associated with grade 3 or 4 liver toxicity, occasionally meeting clinical criteria for VOD (Table 11.1). The latter usually presents with abrupt onset of significant weight gain (more than 5% of pretreatment weight), often associated with ascites, abdominal distension, and right upper quadrant abdominal pain (Giles et al. 2001).

Hepatic transaminase elevation between five to ten times above the normal range was observed in 20% of patients in the phase I study (Sievers et al. 1999). Among 277 patients treated in the phase II studies, grade 3 or 4 hyperbilirubinemia or transaminitis occurred in 38%, with VOD criteria being met in seven patients (3%) and death occurring in four patients (1%) (Sievers et al. 2001).

Hepatic toxicity and VOD have subsequently been observed in other studies, especially when GO is used in previously treated patients, and in combination with other chemotherapeutic agents (Table 11.1). In a comprehensive review of a single-center experience with GO used alone or in combination (Giles et al. 2001), 12% of 119 patients with AML, MDS, or blast-phase chronic myeloid leukemia (CML) developed VOD with 86% mortality, though not clearly attributable to VOD. Of note, 9 out of the 14 patients who developed VOD had received prior chemotherapy and only 3 of the 14 received GO as single-agent therapy.

The incidence of VOD may be even higher when GO is used prior to or after stem cell transplantation. Among the 46 patients who subsequently underwent SCT in the phase II trials (Sievers et al. 2002), VOD was diagnosed in eight (17%) with a mortality of 65% (five out of eight). Similarly, when used as induction therapy prior to SCT, liver toxicity was documented in 11 of 23 patients (48%) with nine deaths (82% mortality) in a single-institution trial (Rajvanshi et al. 2002). This trend was confirmed in a comparative study (Wadleigh et al. 2003), where patients treated with GO prior to allogeneic SCT were found to be more likely to develop VOD than patients transplanted without prior GO treatment (adjusted OR = 21.6). In that study, 9 out of 14 (64%) with GO exposure developed VOD, compared to 4 out of 48 (8%) without GO exposure. In our experience, VOD was diagnosed in one out of eight such patients (12.5%), treated after SCT (Cohen et al. 2002).

The sinusoidal endothelial cell (SEC) appears to play a key role in the development of clinical VOD (Bearman 1995). However, the CD33 antigen was not detectable on unstimulated SECs or SECs activated by various cytokines *in vitro* (Favaloro 1993). Thus, it is unlikely that SEC toxicity is a consequence of direct GO uptake by these cells, especially since VOD has not been reported in association with other anti-CD33 monoclonal antibody-based therapies (Caron et al. 1998; Jurcic et al. 2000; Kossman et al. 1999). Rather, it is far more likely that free calicheamicin, which is extremely toxic upon direct administration without a vehicle, is the initiator of this syndrome (Giles et al. 2001). Free, as well as conjugated, calicheamicin was noted to cause liver toxicity in preclinical testing of GO (Bross et al. 2001). Moreover, hepatic transaminase elevations were seen in the setting of another calicheamicin-based therapy with CMB-401 in the treatment of ovarian malignancy (Gillespie et al. 2000).

Second-generation trials

Relapsed/refractory disease

Based on its nonoverlapping toxicity with the more traditional induction regimens available to patients with AML, GO has been combined with anthracyclines and/or nucleotide analogues with promising results in single-institution trials (Alvarado et al. 2003; Cortes et al. 2002). Incidence of VOD appears to be higher in some of these combination trials, than in the single-agent studies reported previously, ranging from 6% to 14% when GO is given at full dose (Table 11.1).

On the other hand, a randomized phase II trial of the ECOG comparing GO (at a 6 mg/m^2 single dose) and intermediate-dose ara-C (IDAC) vs.

liposomal daunorubicin and IDAC vs. CAT (cyclophosphamide, ara-C, topotecan) in patients with relapsed or refractory AML was closed early because of poor efficacy (Litzow et al. 2003).

De novo disease

A single-institution study of 51 patients aged 65 and over with untreated AML (73%) or high-risk MDS (27%) and unfavorable cytogenetics reported a similar CR rate of 22%, using GO ± IL-11 and intensive consolidation therapy (Estey et al. 2002a).

The European Organization for the Research and Treatment of Cancer (EORTC) Leukemia Group and Gruppo Italiano Malattie Ematologiche dell' Adulto (GIMEMA) performed a phase II study in 57 previously untreated patients aged 61 to 75, with AML, 25% of which was secondary, using MICE (mitoxantrone, ara-C, etoposide) consolidation after single-agent GO (Amadori et al. 2002). The OR rate among those higher-risk untreated patients was 35% (22.8% CR and 12.3% CRp). After treatment with MICE the OR rate increased to 54.4% (35.1% CR and 19.3% CRp). The 1-year overall survival was 30%. A phase III trial is underway in Europe with the GO dose reduced to 6 mg/m^2.

There are several other international cooperative group trials ongoing at this time (Table 11.1). GO (6 mg/m^2 on day 1 and 4 mg/m^2 on day 8) has been combined with standard 7-day infusional ara-C, in patients 60 years of age or older with de novo CD33+ AML (Baccarani et al. 2002). Dose-limiting toxicity in the phase I study was VOD. Four out of seven patients treated thus far in the phase II part have achieved a CR. One out of 12 patients developed grade 4 transaminitis that has resolved uneventfully (Baccarani et al. abst. 1322).

Another ongoing phase I/II study combines GO (6 mg/m^2 on day 4) with daunorubicin and infusional ara-C, in patients younger than 60 years (De Angelo et al. 2003). An additional course of induction with daunorubicin and ara-C was allowed on day 15. Of 43 evaluable patients with de novo AML, 36 achieved a CR (84%) in the phase II part. Grade 3 or 4 transaminitis or hyperbilirubinemia occurred in 67%, without any cases of VOD. However, four of nine patients who went on to receive SCT after induction therapy developed VOD. A Southwestern Oncology Group (SWOG) phase III trial comparing this regimen to standard daunorubicin and ara-C is underway.

In a pilot study of the Medical Research Council (MRC) in Britain, GO (3 mg/m^2 on day 1) was combined with DAT (daunorubicin, ara-C, thioguanine), or DA (daunorubicin, ara-C), or FLAG-Ida (fludarabine, ara-C, GCSF, idarubicin) in 55 patients with de novo AML (Kell et al. 2002). GO was also given in consolidation with MACE (amsacrine, ara-C, etoposide), or high-dose ara-C among responders. Overall, there was a CR rate of 85% but at a price of severe liver toxicity, with 56% of those patients receiving thioguanine developing grade 3 or 4 hepatotoxicity. Trial AML-15 will compare DA ± GO to FLAG-Ida ± GO as induction and MACE ± GO to ara-C ± GO as consolidation.

Multidrug resistance phenotype

P-glycoprotein, the product of the multidrug resistance gene MDR1, is a transmembrane channel implicated in active drug efflux from the cytosol and chemotherapy resistance. Its expression by myeloblasts has been associated with inferior outcomes in AML (Damiani et al. 1998; Legrand et al. 1998; Leith et al. 1997; Leith et al. 1999; Willman 1997). Additionally, the activity of GO in leukemic cell lines is attenuated by the expression of MDR1 and can be restored by MDR1 modification (Linenberger et al. 2001; Matsui et al. 2002; Naito et al. 2000; Walter et al. 2003). It has been shown in the phase I and II studies that leukemia response is correlated with a low capacity of leukemic blasts to extrude 3, 3′-diethyloxacarbocyanine iodide ($DiOC_2$) from their cytoplasm, a measure of functional drug efflux from the cell (Sievers et al. 1999; Sievers et al. 2001). Using cyclosporine A, a known P-glycoprotein transport inhibitor, a group of investigators was able to demonstrate that a significant proportion of $DiOC_2$ efflux in these patients is MDR1 mediated (Linenberger et al. 2001). Specifically, among phase II patients, 72% of those without a marrow response compared to 29% of responders, had baseline $DiOC_2$ efflux inhibited by cyclosporine and thus associated with a P-glycoprotein dependent transport mechanism. Similarly, 52% of nonresponders compared to 24% of responders had baseline $DiOC_2$ efflux inhibited by cyclosporine. Moreover, treatment with cyclosporine *in vitro*, increased the ability of GO to induce apoptosis in leukemic blasts (Linenberger et al. 2001). Based on these results, a single-institution trial has incorporated cyclosporine A along with GO in their AML induction regimen with favorable early results (Tsimberidou et al. 2003).

Future directions

Single-agent GO appears to induce adequate second response rates in patients with relapsed or refractory AML, but its activity in phase II trials remains lower than that of current chemotherapeutic regimens. Its use as single agent appears reasonable in older patients, where its activity is comparable to that of more traditional agents with a more favorable toxicity profile. Given its nonoverlapping toxicity with the standard chemotherapeutic agents, its use in combination regimens for younger patients is being explored in phase III clinical trials. It remains to be seen whether increased efficacy in this setting will come at the cost of increased incidence of hepato-toxicity, which was observed in the trials discussed here.

Aside from combination with other agents, alternative dosing regimens of GO should be investigated further. There is evidence to suggest that CD33 density is associated with tumor cell-kill and that CD33 is continuously shuttled to the surface where it is available to bind antibody (van der Velden et al. 2001). Therefore, increasing antibody dose or shortening dosing inter-vals might improve efficacy dramatically. There is clinical evidence of increased antibody efficacy with hypersaturating doses of another anti-CD33 antibody, HuM195, in patients with AML (Caron et al. 1998). The data on

MDR1 and responsiveness to GO also warrant combination clinical trials with P-glycoprotein activity modulators (e.g., cyclosporine A, verapamil, or valspodar). Following induction, the role of GO in consolidation and maintenance therapy, needs to be examined further in the context of current practice.

Finally, a single-institution trial combining GO with all-trans retinoic acid (ATRA) ± idarubicin in patients with de novo acute promyelocytic leukemia showed a sustained CR rate of 80% (Estey et al. 2002a; Estey et al. 2002b). We thus expect GO to be more extensively studied in this and other myeloid malignancies, including chronic myelogenous leukemia and the myelodysplastic syndromes.

References

Alvarado, Y., Tsimberidou, A., Kantarjian, H., Cortes, J., Garcia-Manero, G., Faderl, S., Thomas, D., Estey, E., Giles, F.J. (2003) "Pilot study of Mylotarg, idarubicin and cytarabine combination regimen in patients with primary resistant or relapsed acute myeloid leukemia," *Cancer Chemotherapy Pharmacology*, 51: 87–90.

Amadori, S., Willemze, R., Suciu, S., Mandelli, F., Sellestag, D., Stauder, R., Ho, A.D., Denzlinger, C., Leone, G., Fabris, P., Muus, P., Feingold, J., Beeldens, F., Anak, O., De Witte, T.M. (2002) "Feasibility, safety and efficacy of gemtuzumab ozogamicin followed by intensive chemotherapy for remission induction in previously untreated patients with AML aged 61–75 years: update of a phase II trial of the EORTC Leukemia Group and GIMEMA," *Blood*, 100: Abst. 1318.

American Cancer Society. (2003) *Cancer Facts and Figures 2003*, Atlanta: ACS, Inc.

Amico, D., Barbui, A.M., Erba, E., Rambaldi, A., Introna, M., Golay, J. (2003) "Differential response of human acute myeloid leukemia cells to gemtuzumab ozogamicin *in vitro*: role of Chk1 and Chk2 phosphorylation and caspase 3," *Blood*, 101: 4589–97.

Andrews, R.G., Singer, J.W., Bernstein, I.D. (1989) "Precursors of colony-forming cells in humans can be distinguished from colony-forming cells by expression of the CD33 and CD34 antigens and light scatter properties," *Journal of Experimental Medicine*, 169: 1721–31.

Andrews, R.G., Takahashi, M., Segal, G.M., Powell, J.S., Bernstein, I.D., Singer, J.W. (1986) "The L4F3 antigen is expressed by unipotent and multipotent colony-forming cells but not by their precursors," *Blood*, 68: 1030–5.

Andrews, R.G., Torok-Storb, B., and Bernstein, I.D. (1983) "Myeloid-associated differentiation antigens on stem cells their progeny identified by monoclonal antibodies," *Blood*, 62: 124–32.

Anonymous (1998) "A systemic collaborative overview of randomized trials comparing idarubicin with daunorubicin (or other anthracyclines) as induction therapy for acute myeloid leukaemia. The AML Collaborative Group," *British Journal of Haematology*, 103: 100–9.

Appelbaum, F.R., Matthews, D.C., Eary, J.F., Badger, C.C., Kellogg, M., Press, O.W., Martin, P.J., Fisher, D.R., Nelp, W.B., Thomas, E.D., Bernstein, I.D. (1992) "The use of radiolabeled anti-CD33 antibody to augment marrow irradiation prior to marrow transplantation for acute myelogenous leukemia," *Transplantation*, 54: 829–33.

Archimbaud, E., Thomas, X., Leblond, V., Michallet, M., Fenaux, P., Cordonnier, C., Dreyfus, F., Troussard, X., Jaubert, J., Travade, P., Trancy, J., Assouline, D., Fiere, D. (1995) "Time sequential chemotherapy for previously treated patients with acute myeloid leukemia: long-term follow-up of the etoposide, mitoxantrone, and cytarabine-86 trial," *Journal of Clinical Oncology*, 13: 11–18.

Arlin, Z., Case, D.C. Jr, Moore, J., Wiernik, P., Feldman, E., Saletan, S., Desai, P., Sia, L., Cartwright, K. (1990) "Randomized multicenter trial of cytosine arabinoside with mitoxantrone or daunorubicin in previously untreated adult patients with acute nonlymphocytic leukemia (ANLL). Lederle Cooperative Group," *Leukemia*, 3: 177–83.

Baccarani, M., Durrant, S., Linkesh, W., Lechner, K., Cooper, M., Coutre, S., et al. (2002) "Preliminary report of a phase 2 study of the safety and efficacy of Gemtuzumab Ozogamicin (Mylotarg) given in combination with cytarabine in patients with acute myeloid leukemia," *Blood*, 100(11): Abs 1322.

Ball, E.D., Bernier, G.M., Cornwell, G.G. (1983) "Monoclonal antibodies to myeloid differentiation antigens: *in vivo* studies of three patients with acute myelogenous leukemia," *Blood*, 62: 1203–10.

Barclay, K.L., Yeong, M.L. (1994) "Actinomycin D associated hepatic veno-occlusive disease — a report of 2 cases," *Pathology*, 26: 257–60.

Baudard, M., Marie, J.P., Cadiou, M., Viguie, F., Zittoun, R. (1994) "Acute myelogenous leukaemia in the elderly: retrospective study of 235 consecutive patients," *British Journal of Haematology*, 86(1): 82–91.

Bearman, S.I. (1995) "The syndrome of hepatic veno-occlusive disease after marrow transplantation," *Blood*, 85: 3005–20.

Berman, E., Heller, G., Santorsa, J., McKenzie, S., Gee, T., Kempin, S., Gulati, S., Andreeff, M., Kolitz, J., Gabrilove, J., et al. (1991) "Results of a randomized trial comparing idarubicin and cytosine arabinoside with daunorubicin and cytosine arabinoside in adult patients with newly diagnosed acute myelogenous leukemia," *Blood*, 77: 1666–74.

Bernstein, I.D., Singer, J.W., Andrews, R.G., Keating, A., Powell, J.S., Bjomson, B.H., Cuttner, J., Najfeld, V., Reaman, G., Raskind, W., et al. (1987) "Treatment of acute myeloid leukemia cells *in vitro* with monoclonal antibody recognizing a myeloid differentiation antigen allows normal progenitor cells to be expressed," *Journal of Clinical Investigation*, 79: 1153–9.

Bernstein, I.D., Tam, M.R., Nowinski, R.C. (1980) "Mouse leukemia: therapy with monoclonal antibodies against a thymus differentiation antigen," *Science*, 207: 68–71.

Bross, P.F., Beitz, J., Chen, G., Chen, X.H., Duffy, E., Kieffer, L., Roy, S., Sridhara, R., Rahman, A., Williams, G., Pazdur, R. (2001) "Approval summary: gemtuzumab ozogamicin in relpased acute myeloid leukemia," *Clinical Cancer Research*, 7: 1490–5.

Candoni, A., Damiani, D., Michelutti, A., Masolini, P., Michieli, M., Michelutti, T., Geromin, A., Fanin, R. (2003) "Clinical characteristics, prognostic factors and multidrug-resistance related protein expression in 36 adult patients with acute promyelocytic leukemia," *European Journal of Haematology*, 71: 1–8.

Caron, P.C., Dumont, L., Scheinberg, D.A. (1998) "Supersaturating infusional humanized anti-CD33 monoclonal antibody HuM195 in myelogenous leukemia," *Clinical Cancer Research*, 4: 1421–8.

Cassileth, P.A., Harrington, D.P., Appelbaum, F.R., Lazarus, H.M., Rowe, J.M., Paietta, E., Willman, C., Hurd, D.D., Bennett, J.M., Blume, K.G., Head, D.R., Wiernik, P.H. (1998) "Chemotherapy compared with autologous or allogeneic bone marrow transplantation in the management of acute myeloid leukemia in first remission," *New England Journal of Medicine*, 339: 1649–56.

Clift, R.A., Buckner, C.D., Thomas, E.D., Kopecky, K.J., Appelbaum, F.R., Tallman, M., Storb, R., Sanders, J., Sullivan, K., Banaji, M., et al. (1987) "The treatment of acute non-lymphoblastic leukemia by allogeneic marrow transplantation," *Bone Marrow Transplantation*, 2: 243–58.

Cohen, A.D., Luger, S.M., Sickles, C., Mangan, P.A., Porter, D.L., Schuster, S.J., Tsai, D.E., Nasta, S., Gewirtz, A.M., Stadtmauer, E.A. (2002) "Gemtuzumab ozogamicin (Mylotarg) monotherapy for relapsed AML after hematopoietic stem cell transplant: efficacy and incidence of hepatic veno-occlusive disease," *Bone Marrow Transplantation*, 30: 23–8.

Cortes, J., Tsimberidou, A.M., Alvarez, R., Thomas, D., Beran, M., Kantarjian, H., Estey, E., Giles, F.J. (2002) "Mylotarg combined with topotecan and cytarabine in patients with refractory acute myelogenous leukemia," *Cancer Chemotherapy and Pharmacology*, 50: 497–500.

Damiani, D., Michieli, M., Ermacora, A., Candoni, A., Raspadori, D., Geromin, A., Stocchi, R., Grimaz, S., Masolini, P., Michelutti, A., Scheper, R.J., Baccarani, M., et al. (1998) "P-glycoprotein (PGP), and not lung resistance-related protein (LRP), is a negative prognostic factor in secondary leukemias," *Haematologica*, 83: 290–7.

Davis, C.L., Rohatiner, A.Z., Lim, J., Whelan, J.S., Oza, A.M., Amess, J., Love, S., Stead, E., Lister, T.A. (1993) "The management of recurrent acute myelogenous leukemia at a single center over a fifteen-year period," *British Journal of Haematology*, 83: 404–11.

De Angelo, D.J., Schiffer, C., Stone, R., Amrein, P., Fernandez, H., Bradtsock, K., Tallman, M., Foran, J., Juliusson, G., Liu, D., Paul, C., Russo, D., Senke, L., Leopold, L., Stevenson, D., Ritchie, M., Berger, M. (2003) "Preliminary report of a phase 2 study of gemtuzumab ozogamicin in combination with cytarabine and daunorubicin in patients < 60 years of age with de novo acute myeloid leukemia," *American Society of Clinical Oncology Annual Meeting*, Abst. 2325. Online. Available at <http://www.asco.org/ac/1,1003,_12-002489-00_18-002003-00_19-00101382,00.asp> (July 28, 2003).

Dillman, R.O., Shawler, D.L., Dillman, J.B., Royston, I. (1984) "Therapy of chronic lymphocytic leukemia and cutaneous T-cell lymphoma with T101 monoclonal antibody," *Journal of Clinical Oncology*, 2: 881–91.

Dinndorf, P.A., Andrews, R.G., Benjamin, D., Ridgway, D., Wolff, L., Bernstein, I.D. (1986) "Expression of normal myeloid-associated antigens by acute leukemia Cells," *Blood*, 67: 1048–53.

Dowell, J.A., Korth-Bradley, J., Liu, H., King, S.P., Berger, M.S. (2001) "Pharmacokinetics of gemtuzumab ozogamicin, an antibody-targeted chemotherapy agent for the treatment of acute myeloid leukemia in first relapse," *Journal of Clinical Pharmacology*, 41: 1206–14.

Estey, E.H., Cortes, J.E., Garcia-Manero, G., Thomas, D., Faderl, S.H., Verstovsek, S., Pierce, S., O'Brien, S., Beran, M., Kantarjian, H.M., Giles, F.J. (2002c) "Further observations on use of gemtuzumab Ozogamycin (GO) and all-trans retinoic acid (ATRA) in untreated APL: median additional 8 months of follow-up," *Blood*, 100: Abst. 1319.

Estey, E.H., Giles, F.J., Beran, M., O'Brien, S., Pierce, S.A., Faderl, S.H., Cortes, J.E., Kantarjian, H.M. (2002b) "Experience with gemtuzumab ozogamycin ('mylotarg') and all-trans retinoic acid in untreated acute promyelocytic leukemia," *Blood*, 99: 4222–4.

Estey, E.H., Thall, P.F., Giles, F.J., Wang, X.M., Cortes, J.E., Beran, M., Pierce, S.A., Thomas, D.A., Kantarjian, H.M. (2002a) "Gemtuzumab ozogamicin with or without interleukin 11 in patients 65 years of age or older with untreated acute myeloid leukemia and high-risk myelodysplastic syndrome: comparison with idarubicin plus continuous-infusion, high-dose cytosine arabinoside," *Blood*, 99: 4343–9.

Favaloro, E. (1993) "Differential expression of surface antigens on activated endothelium," *Immunology and Cell Biology*, 71: 571–81.

Foon, K.A., Schroff, R.W., Bunn, P.A., Mayer, D., Abrams, P.G., Fer, M., Ochs, J., Bottino, G.C., Sherwin, S.A., Carlo, D.J., et al. (1984) "Effects of monoclonal antibody therapy in patients with chronic lymphocytic leukemia," *Blood*, 64: 1085–93.

Freeman, S.D., Kelm, S., Barber, E.K., Crocker, P.R. (1995) "Characterization of CD33 as a new member of the sialoadhesin family of cellular interaction molecules," *Blood*, 85: 2005–12.

Gale, R.P., Horowitz, M.M., Rees, J.K., Gray, R.G., Oken, M.M., Estey, E.H., Kim, K.M., Zhang, M.J., Ash, R.C., Atkinson, K., Champlin, R.E., Dicke, K.A., Gajewski, J.L., Goldman, J.M., Helbig, W., Henslee-Downey, P.J., Hinterberger, W., Jacobsen, N., Keating, A., Klein, J.P., Marmont, A.M., Prentice, H.G., Reiffers, J., Rimm, A.A., Rowlings, P.A., Sobocinski, K.A., Speck, B., Wingard, J.R., Bortin, M.M. (1996) "Chemotherapy versus transplants for acute myelogenous leukemia in second remission," *Leukemia*, 10: 13–19.

Geller, R.B., Saral, R., Piantadosi, S., Zahurak, M., Vogelsang, G.B., Wingard, J.R., Ambinder, R.F., Beschorner, W.B., Braine, H.G., Burns, W.H., et al. (1989) "Allogeneic bone marrow transplantation after high-dose busulfan and cyclophosphamide in patients with acute nonlymphocytic leukemia," *Blood*, 73: 2209–18.

Giles, F.J., Kantarjian, H.M., Kornblau, S.M., Thomas, D.A., Garcia-Manero, G., Waddelow, T.A., David, C.L., Phan, A.T., Colburn, D.E., Rashid, A., Estey, E.H. (2001) "Mylotarg (gemtuzumab ozogamicin) therapy is associated with hepatic venoocclusive disease in patients who have not received stem cell transplantation," *Cancer*, 92: 406–13.

Gill, R.A., Onstad, G.R., Cardamone, J.M., Maneval, D.C., Summer, H.W. (1982) "Hepatic veno–occlusive disease caused by 6-thioguanine," *Annals of Internal Medicine*, 96: 58–60.

Gillespie, A.M., Broadhead, T.J., Chan, S.Y. Owen, J., Farnsworth, A.P., Sopwith, M., Coleman, R.E. (2000) "Phase I open study of the effects of ascending doses of the cytotoxic immunoconjugate CMB-401 (hCTMO1-calicheamicin) in patients with epithelial ovarian cancer," *Annals of Oncology*, 11: 735–41.

Griffin, J.D., Linch, D., Sabbath, K., Larcom, P., Schlossman, S.F. (1984) "A monoclonal antibody reactive with normal and leukemic human myeloid progenitor cells," *Leukemia Research*, 8: 521–34.

Hamann, P.R., Hinman, L.M., Beyer, C.F., et al. (2002a) "An anti-CD33 antibody-calicheamicin conjugate for treatment of acute myeloid leukemia: choice of linker," *Bioconjugate Chemistry*, 13: 40–6.

Hamann, P.R., Hinman, L.M., Hollander, I., et al. (2002b) "Gemtuzumab ozogamicin, a potent and selective anti-CD33 antibody-calicheamicin conjugate for treatment of acute myeloid leukemia," *Bioconjugate Chemistry*, 13: 47–58.

Hiddemann, W., Kern, W., Schoch, C., Fonatsch, C., Heinecke, A., Vormann, B., Buchner, T. (1999) "Management of acute myeloid leukemia in elderly patients," *Journal of Clinical Oncology*, 17: 3569–76.

Hinman, L.M., Hamann, P.R., Wallace, R., Menendez, A.T., Durr, F.E., Upeslacis, J. (1998) "Preparation and characterization of monoclonal antibody conjugates of the calicheamicins: a novel and potent family of antitumor antibiotics," *Cancer Research*, 53: 3336–42.

Ikemoto, N., Kumar, R.A., Ling, T.T., Danishefsky, S.J., and Patel, D.J. (1995) "Calicheamicin-DNA compexes: warhead alignment and saccharide recognition of the minor groove," *Proceedings of the National Academy of Sciences USA*, 92: 10506–10.

Jilani, I., Estey, E., Huh, Y., Joe, Y., Manshouri, T., Yared, M., Giles, F., Kantarjian, H., Cortes, J., Thomas, D., Keating, M., Freireich, E., Albitar, M. (2002) "Differences in CD33 intensity between various myeloid neoplasms," *American Journal of Clinical Pathology*, 118: 560–6.

Jones, R.J., Lee, K.S., Beschorner, W.E., Vogel, V.G., Grochow, L.B., Braine, H.G., Vogelsang, G.B., Sensenbrenner, L.L., Santos, G.W., Saral, R. (1987) "Venoocclusive disease of the liver following bone marrow transplantation," *Transplantation*, 44: 778–83.

Jurcic, J.G., DeBlasio, T., Dumont, L., Yao T.J., and Scheinberg, D.A. (2000) "Molecular remission induction with retinoic acid and anti–CD33 Monoclonal antibody HuM195 in acute promyelocytic leukemia," *Clinical Cancer Research*, 6(2): 372–80.

Jurcic, J.G., Larson, S.M., Sgouros, G., McDevitt, M.R., Finn, R.D., Divgi, C.R., Ballangrud, A.M., Hamacher, K.A., Ma, D., Humm, J.L., Brechbiel, M.W., Molinet, R., Scheinberg, D.A. (2002) "Targeted alpha particle immunotherapy for myeloid leukemia," *Blood*, 100: 1233–9.

Jurcic, J.G., Philip, C.C., Nikura, T.K., Paradopoulos, E.B., Finn, R.D., Gansow, O.A., Miller, W.H., Jr, Geerlings, M.W., Warrell, R.P., Jr, Larson, S.M., Scheinberg, D.A. (1995) "Radiolabeled anti-CD33 monoclonal antibody M195 for myeloid leukemias," *Cancer Research*, 55: 5908s–10s.

Keating, M.J., Kantarjian, H., Smith, T.L., Estey, E., Walters, R., Andersson, B., Beran, M., McCredie, K.B., Freireich, E.J. (1989) "Response to salvage therapy and survival after relapse in acute myelogenous leukemia," *Journal of Clinical Oncology*, 7: 1071–80.

Kell, J.W., Burnett, A.K., Chopra, R., Yin, J., Culligan, D., Clark, R., et al. (2002) "Mylotarg (Gemtuzumab Ozogamicin: GO) given simultaneously with intensive induction and/or consolidation therapy for AML is feasible and may improve the response rate," *Blood*, 100(11): Abs 746.

Kern, W., Aul, C., Maschmayer, G., Schonrock-Nabulsi, R., Ludwig, W.D., Bartholomaus, A., Bettelheim, P., Wormann, B., Buchner, T., Hidemann, W. (1998) "Superiority of high-dose over intermediate-dose cytosine arabinoside in the treatment of patients with high-risk acute myeloid leukemia: results of an age-adjusted prospective randomized comparison," *Leukemia*, 12: 1049–55.

Kirch, M.E., Hammerling, U. (1981) "Immunotherapy of murine leukemias by monoclonal antibody: Effect of passively administered antibody on growth of transplanted tumor cells," *Journal of Immunology*, 127: 805–10.

Korth-Bradley, J.M., Dowel, J.A., King, S.P., Liu, H., Berger, M.S. Mylotarg Study Group. (2001) "Impact of age and gender on the pharmacokinetics of Gemtuzumab ozogamicin," *Pharmacotherapy*, 21: 1175–80.

Kossman, S.E., Scheinberg, D.A., Jurcic, J.G., Jimenez, J., Caron, P.C. (1999) "A phase I trial of humanized monoclonal antibody HuM195 (anti-CD33) with low-dose interleukin 2 in acute myelogenous leukemia," *Clinical Cancer Research*, 5: 2748–55.

Larson, R.A., Boogaerts, M., Estey, E., Karanes, C., Stadtmauer, E.A., Sievers, E.L., Mineur, P., Bennett, J.M., Berger, M.S., Eten, C.B., Munteanu, M., Loken, M.R., Van Dongen, J.J., Bernstein, I.D., Appelbaum, F.R. Mylotarg Study Group. (2002b) "Antibody-targeted chemotherapy of older patients with acute myeloid leukemia in first relapse using Mylotarg (gemtuzumab ozogamicin)," *Leukemia*, 16: 1627–36.

Larson, R.A., Sievers, E.L., Stadtmauer, E.A., Löwenberg, B., Estey, E., Dombret, H., Theobald, M., Voliotis, D., Leopold, L.H., Richie, M., Berger, M.S., Sherman, M.L., Appelbaum, F.R. Mylotarg Study Group. (2002a) "A final analysis of the efficacy and safety of gemtuzumab ozogamicin in 277 patients with acute myeloid leukemia in first relapse," *Blood*, 100: Abst. 1312.

Lee, M.D., Dunne, T.S., Chang, C.C., Ellestad, G.A., Siegel, M.M., Morton, G.O., McGahren, W.J., Borders, D.B. (1987b) "Calicheamicins, a novel family of antitumor antibiotics. 2: chemistry and structure of calicheamicin g1 I," *Journal of the American Chemical Society*, 109: 3466–8.

Lee, M.D., Dunne, T.S., Siegel, M.M., Chang, C.C., Morton, G.O., Borders, D.B. (1987a) "Calicheamicins, a novel family of antitumor antibiotics. 1: Chemistry and partial structure of calicheamicin g1 I," *Journal of the American Chemical Society*, 109: 3464–6.

Legrand, O., Perrot, J.Y., Baudard, M., Cordier, A., Lautier, R., Simonin, G., Zittoun, R., Casadevall, N., Marie, J.P. (2000) "The immunophenotype of 177 adults with acute myeloid leukemia: proposal of a prognostic score," *Blood*, 96: 870–7.

Legrand, O., Simonin, G., Perrot, J.Y., Zittoun, R., Marie, J.P. (1998) "Pgp and MRP activities using calcein-AM are prognostic factors in adult acute myeloid leukemia patients," *Blood*, 91: 4480–8.

Leith, C.P., Kopecky, K.J., Chen, I.M., Ejidems, L., Slovak, M.L., McConnell, T.S., Head, D.R., Weick, J., Grever, M.R., Appelbaum, F.R., Willman, C.L. (1999) "Frequency and clinical significance of the expression of the multidrug resistance proteins MDR1/P-glycoprotein, MRP1, and LRP in acute myeloid leukemia: a Southwest Oncology Group study," *Blood*, 94: 1086–99.

Leith, C.P., Kopecky, K.J., Godwin, J., McConnell, T., Slovak, M.L., Chen, I.M., Head, D.R., Appelbaum, F.R., Willman, C.L. (1997) "Acute myeloid leukemia in the elderly: assessment of multidrug resistance (MDR1) and cytogenetics distinguishes biologic subgroups with remarkably distinct responses to standard chemotherapy. A Southwest Oncology Group Study," *Blood*, 89(9): 3323–9.

Lemoli, R.M., Casparetto, C., Scheinberg, D.A., Moore, M.A., Clarkson, B.D., Gulati, S.C. (1991) "Autologous bone marrow transplantation in acute myelogenous leukemia: *in vitro* treatment with myeloid-specific monoclonal antibodies and drugs in combination," *Blood*, 77: 1829–36.

Linenberger, M.L., Hong, T., Flowers, D., Sievers, E.L., Gooley, T.A., Bennett, J.M., Berger, M.S., Leopold, L.H., Appelbaum, F.R., Bernstein, I.D. (2001) "Multidrug-resistance phenotype and clinical responses to gemtuzumab ozogamicin," *Blood*, 98: 988–94.

Litzow, M.R., Goloubeva, O., Rowe, J., Cripe, L., Tallman, M., Gore, S. (2003) "A randomized phase II trial of gemtuzumab ozogamicin vs. liposomal daunorubicin vs. cyclophosphamide plus topotecan, each combined with intermediate dose Ara-C for the treatment of patients with relapsed or refractory acute myelogenous leukemia," *American Society of Clinical Oncology Annual Meeting*, Abst. 2359. Online. Available at <http://www.asco.org/ac/1,1003,_12-002489-00_18-002003-00_19-00100326,00.asp> (July 28, 2003).

Löwenberg, B., Downing, J.R., and Burnett, A. (1999) "Acute myeloid leukemia," *New England Journal of Medicine*, 341: 1051–61.

Löwenberg, B., Suciu, S., Archimbaud, E., Haak, H., Stryckmans, P., de Cataldo, R., Dekker, A.W., Berneman, Z.N., Thyss, A., van der Lelie, J., Sonneveld, P., Visani, G., Fillet, G., Hayat, M., Hagemeijer, A., Solbu, G., Zittoun, R. (1998) "Mitoxantrone versus daunorubicin in induction–consolidation chemotherapy — the value of low-dose cytarabine for maintenance of remission, and an assessment of prognostic factors in acute myeloid leukemia in the elderly: final report. European Organization for the Research and Treatment of Cancer and the Dutch-Belgian Hemato-Oncology Cooperative Hovon Group," *Journal of Clinical Oncology*, 16: 872–81.

Mandelli, F., Petti, M.C., Ardia, A., Di Pietro, N., Di Raimondo, F., Ganzina, F., Falconi, E., Geraci, E., Ladogana, S., Latagliata, R., Malleo, C., Nobile, F., Petti, N., Rotoli, B., Specchia, G., Tabilio, A., Resegotti, L. (1991) "A randomised clinical trial comparing idarubicin and cytarabine to daunorubicin and cytarabine in the treatment of acute non-lymphoid leukemia: a multicentric study from the Italian Co-operative Group GIMEMA," *European Journal of Cancer*, 27: 750–5.

Matsui, H., Takeshita, A., Naito, K., Shinjo, K., Shigeno, K., Maekawa, M., Yamakawa, Y., Tanimoto, M., Kobayashi, M., Ohnishi, K., Ohno, R. (2002) "Reduced effect of gemtuzumab ozogamicin (CMA-676) on P-glycoprotein and/or CD34-positive leukemia cells and its restoration by multidrug resistance modifiers," *Leukemia*, 16: 813–9.

Mayer, R.J., David, R.B., Schiffer, C.A., Berg, D.T., Powell, B.L., Schulman, P., Omura, G.A., Moore, J.O., McIntyre, O.R., Frei, E. (1994) "Intensive postremission chemotherapy in adults with acute myeloid leukemia," *New England Journal of Medicine*, 331: 896–903.

McDonald, G.B., Hinds, M.S., Fisher, L.D., Schoch, H.G., Wolford, J.L., Banaji, M., et al. (1993) "Veno-occlusive disease of the liver and multiorgan failure after bone marrow transplantation: a cohort study of 355 patients," *Annals of Internal Medicine*, 118: 255–67.

Nadler, L.M., Stashenko, P., Hardy, R., Kaplan, W.D., Button, L.N., Kufe, D.W., Antman, K.H., Schlossman, S.F. (1980) "Serotherapy of a patient with a monoclonal antibody directed against a human lymphoma-associated antigen," *Cancer Research*, 40: 3147–54.

Naito, K., Takeshita, A., Shigeno, K., Nakamura, S., Fujisawa, S., Shinjo, K., Yoshida, H., Ohnishi, K., Mori, M., Terakawa, S., Ohno, R. (2000) "Calicheamicin-conjugated humanized anti-CD33 monoclonal antibody (gemtuzumab ozogamicin, CMA-676) shows cytocidal effect on CD33-positive leukemia cell lines, but is inactive on P-glycoprotein-expressing sublines," *Leukemia*, 14: 1436–43.

Nicolaou, K.C., Smith, A.L., Yue, E.W. (1993) "Chemistry and biology of natural and designed enediynes," *Proceedings of the National Academy of Sciences USA*, 90: 5881–8.

Ortega, J.A., Donaldsonm S.S., Ivym S.P., et al. (1997) "Venoocclusive disease of the liver after chemotherapy with vincristine, actinomycin D, and cyclophosphamide for the treatment of rhabdomyosarcoma," *Cancer*, 79: 2435–9.

Pagliaro, L.C., Liu, B., Munker, R., Andreeff, M., Freireich, E.J., Scheinberg, D.A., Rosenblum, M.G. (1998) "Humanized M195 monoclonal antibody conjugated to recombinant gelonin: an anti-CD33 immunotoxin with antileukemic activity," *Clinical Cancer Research*, 4: 1971–6.

Peiper, S.C., Ashmun, R.A., Look, A.T. (1988) "Molecular cloning, expression, and chromosomal localization of a human gene encoding the CD33 myeloid differentiation antigen," *Blood*, 72: 314–21.

Pierelli, L., Teofili, L., Menichella, G., Rumi, C., Paoloni, A., Iovino, S., Puggioni, P.L., Leone, G., Bizzi, B. (1993) "Further investigations on the expression of HLA-DR, CD33, and CD13 surface antigens in purified bone marrow and peripheral blood CD34+ hematopoietic progenitor cells," *British Journal of Haematology*, 84: 24–30.

Rajvanshi, P., Shulman H.M., Sievers, E.L., McDonald, G.B. (2002) "Hepatic sinusoidal obstruction after gemtuzumab ozogamicin (Mylotarg) therapy," *Blood*, 99: 3915.

Rees, J.K., Gray, R.G., Swirsky, D., Hayhoe, F.G. (1986) "Principal results of the Medical Research Council's 8th acute myeloid leukemia trial," *Lancet*, 2: 1236–41.

Ries, L.A.G., Eisner, M.P., Kosary, C.L., Hankey, B.F., Miller, B.A., Clegg, L., Mariotto, A., Fay, M.P., Feuer, E.J., Edwards, B.K. (eds.). (2003) *SEER Cancer Statistics Review, 1975–2000*, Bethesda, MD: National Cancer Institute. Online. Available at <http://seer.cancer.gov/csr/1975_2000> (July 7, 2003).

Ritz, J., Pesando, J.M., Sallan, S.E., Clavell, L.A., Notis-McConarty, J., Rosenthal, P., Schlossman, S.F. (1981) "Serotherapy of acute lymphoblastic leukemia with monoclonal antibody," *Blood*, 58: 141–52.

Robertson, M.J., Soiffer, R.J., Freedman, A.S., Rabinowe, S.L., Anderson, K.C., Ervin, T.J., Murray, C., Dear, K., Griffin, J.D., Nadler, L.M., et al. (1992) "Human bone marrow depleted of CD33-positive cells mediates delayed but durable reconstitution of hematopoiesis: clinical trial of MY9 monoclonal antibody–purged autografts for the treatment of acute myeloid leukemia," *Blood*, 79: 2229–36.

Rowe, J.M. (2000) "Treatment of acute myelogenous leukemia in older adults," *Leukemia*, 14: 480–7.

Rowley, J.D., Alimena, G., Garson, O.M., Hageneijer, A., Mitelman, F., Prigogina, E.L. (1982) "A collaborative study of the relationship of the morphological type of acute nonlymphocytic leukemia with patients age and karyotype," *Blood*, 59: 1013–32.

Roy, D.C., Griffin, J.D., Belvin, M., Blattler, W.A., Lambert, J.M., Ritz, J. (1991) "Anti-MY9-blocked-ricin: an immunotoxin for selective targeting of acute myeloid leukemia cells," *Blood*, 77: 2404–12.

Sabbath K.D., Ball, E.D., Larcom, P., Davis, R.B., Griffin, J.D. (1985) "Heterogeneity of clonogenic cells in acute myeloblastic leukemia," *Journal of Clinical Investigation*, 75: 746–53.

Scheinberg, D.A., Lovett, D., Divgi, C.R., Graham, M.C., Berman, E., Pentow, K., Feirt, N., Finn, R.O., Clarkson, B.O., Gee, T.S., Larson, S.M., Oettgen, H.F., Old, L.J. (1991) "A phase I trial of monoclonal antibody M195 in acute myelogenous leukaemia: specific bone marrow targeting and internalization of radionuclide," *Journal of Clinical Oncology*, 9: 478–90.

Scheinberg, D.A., Tanimoto, M., McKenzie, S., Strife, A., Old, L.J., Clarkson, B.D. (1989) "Monoclonal antibody M195: a diagnostic marker for acute myelogenous leukemia," *Leukemia*, 3: 440–5.

Schwartz, M.A., Lovett, D.R., Redner, A., Finn, R.D., Graham, M.C., Divgi, C.R., Dantis, L., Gee, T.S., Andreeff, M., Old, L.J. (1993) "Dose-escalation trial of M195 labeled with iodine 131 for cytoreduction and marrow ablation in relapsed or refractory myeloid leukemias," *Journal of Clinical Oncology*, 11: 294–303.

Sievers, E.L., Appelbaum, R.T., Spielberger, R.T., Forman, S.J., Flowers, D., Smith, F.O., Shannon-Dorcy, K., Berger, M.S., Bernstein, I.D. (1999) "Selective ablation of acute myeloid leukemia using antibody-targeted chemotherapy: a phase I study of an anti-CD33 calicheamicin immunoconjugate," *Blood*, 93: 3678–84.

Sievers, E., Larson, R., Estey, E., Löwenberg, B., Leopold, L., Berger, M., Appelbaum, F. Mylotarg Study Group. (2002) "Final report of prolonged disease-free survival in patients with acute myeloid leukemia in first relapse treated with gemtuzumab ozogamicin followed by hematopoietic stem cell transplantation," *Blood*, 100: Abst. 327

Sievers, E.L., Larson, R.A., Stadtmauer, E.A., Estey, E., Löwenberg, B., Dombret, H., Karanes, C., Theobald, M., Bennett, J.M., Sherman, M.L., Berger, M.S., Eten, C.B., Loken, M.R., van Dongen, J.J., Bernstein, I.D., Appelbaum, F.R. Mylotarg Study Group. (2001) "Efficacy and safety of gemtuzumab ozogamicin in patients with CD33-positive acute myeloid leukemia in first relapse," *Journal of Clinical Oncology*, 19: 3244–54.

Stadtmauer, E., Sievers, E., Larson, R., Estey, E., Löwenberg, B., Leopold, L., Berger, M., Appelbaum, F. (2002) "Final report of the effect of cytogenetic risk group on outcome of patients with acute myeloid leukemia in first relapse treated with gemtuzumab ozogamicin (Mylotarg®)," *Blood*, 100: Abst. 740.

Tack, D.K., Letendre, L., Kamath, P.S., Tefferi, A. (2001) "Development of hepatic veno-occlusive disease after Mylotarg infusion for relapsed acute myeloid leukemia [REPORT]," *Bone Marrow Transplantation*, 28(9): 895–7.

Tanimoto, M., Scheinberg, D.A., Cordon-Cardo, C., Huie, D., Clarkson, B.D., and Old, L.J. (1989) "Restricted expression an early myeloid and monocytic cell surface antigen defined by monoclonal antibody M195," *Leukemia*, 3: 339–48.

Taylor, V.C., Buckley, C.D., Douglas, M., Cody, A.J., Simmons, D.L., Freeman, S.D. (1999) "The myeloid-specific sialic acid-binding receptor, CD33, associates with the protein-tyrosine phosphatases, SHP-1 and SHP-2," *Journal of Biological Chemistry*, 274: 11505–12.

Tsimberidou, A., Estey, E., Cortes, J., Thomas, D., Faderl, S., Verstovsek, S., Garcia—Manero, G., Keating, M., Albitar, M., O'Brien, S., Kantarjian, H., Giles, F. (2003) "Gemtuzumab, fludarabine, cytarabine, and cyclosporine in patients with newly diagnosed acute myelogenous leukemia or high-risk myelodysplastic syndromes," *Cancer*, 97: 1481–7.

Ulyanova, T., Blasioli, J., Woodford-Thomas, T.A., Thomas, M.L. (1999) "The sialoadhesin CD33 is a myeloid-specific inhibitory receptor," *European Journal of Immunology*, 29: 3440–9.

van der Jagt, R.H.C., Badger, C.C., Appelbaum, F.R., Press, O.W., Matthews, D.C., Eary, J.F., Krohn, K.A., and Bernstein, I.D. (1992) "Localization of radiolabeled antimyeloid antibodies in a human acute leukemia xenograft tumor model," *Cancer Research*, 52: 89–94.

van der Velden, V.H., te Marvelde, J.G., Hoogeveen, P.G., Bernstein, I.D., Houtsmuller, A.B., Berger, M.S., van Dongen, J.J. (2001) "Targeting of the CD33-calicheamicin immunoconjugate Mylotarg (CMA-676) in acute myeloid leukemia: *in vivo* and *in vitro* saturation and internalization by leukemic and normal myeloid cells," *Blood*, 97: 3197–204.

Vogler, W.R., McCarley, D.L., Stagg, M., Bartolucci, A.A., Moore, J., Martelo, O., Omura, G.A. (1994) "A phase III trial of high-dose cytosine arabinoside with or without etoposide in relapsed and refractory acute myelogenous leukemia. A Southeastern Cancer Study Group trial," *Leukemia*, 8: 1847–53.

Wadleigh, M., Richardson, P.G., Zahrieh, D., Lee, S.J., Cutler, C., Ho, V., Alyea, E.P., Antin, J.H., Stone, R.M., Soiffer, R.J., DeAngelo, D.J. (2003) "Prior gemtuzumab ozogamicin exposure significantly increases the risk of veno-occlusive disease in patients who undergo myeloablative allogeneic stem cell transplantation," *Blood*, Online posting (May 8, 2003).

Wagner, J.E., Collins, D., Fuller, S., Schain, L.R., Berson, A.E., Almici, C., Hall, M.A., Chen, K.E., Okarma, T.B., Lebkowski, J.S. (1995) "Isolation of small, primitive human hematopoietic stem cells: distribution of cell surface cytokine receptors and growth in SCID-Hu mice," *Blood*, 86: 512–23.

Walter, R.B., Raden, B.W., Hong, T.C., Flowers, D.A., Bernstein, I.D., Linenberger, M.L. (2003) "Multidrug resistance protein (MRP) attenuates gemtuzumab ozogamicin-induced cytotoxicity in acute myeloid leukemia cells," *Blood*, Online posting (April 10, 2003).

Webb, D.K.H. (1999) "Management of relapsed acute myeloid leukaemia," *British Journal of Haematology*, 106: 851–9.

Wiernik, P.H., Banks, P.L., Case, D.C., Jr, Arlin, Z.A., Periman, P.O., Todd, M.B., Ritch, P.S., Enck, R.E., Weitberg, A.B. (1992) "Cytarabine plus idarubicin or daunorubicin as induction and consolidation therapy for previously untreated adult patients with acute myeloid leukemia," *Blood*, 79: 313–9.

Willman, C.L. (1997) "The prognostic significance of the expression and function of multidrug resistance transporter proteins in AML: studies of the Southwest Oncology Leukemia Research Program," *Seminars in Hematology*, 34: 25–33.

Zein, N., Poncin, M., Nilakantan, R., Ellestad, G.A. (1989) "Calicheamicin g1 I and DNA: molecular recognition process responsible for site specificity," *Science*, 244: 697–9.

Zein, N., Sinha, A.M., McGahren, W.J., Ellestad, G.A. (1988) "Calicheamicin gamma 1: an antitumor antibiotic that cleaves double-stranded DNA site specifically," *Science*, 240: 1198–201.

Index

  adhesion 88
FLAIR (fluid-attenuated inversion
    recovery) 107
flow cytometry 163
flucanozole 164
fludarabine 127, 162
fluid overload 158
flushing 123
focal adhesion kinase (FAK) 103
Food and Drug Administration (FDA) 33
fungoides 122
fusion
  cellular intoxication by 126
  clearance of 118
  half-time 128
  immunotoxin 117
  intoxication by 154
  ligand 119
  proteins 173, 179
  toxin 120, 128

galactose 75, 76
gastritis 190, 191, 193
gastrointestinal ulceration 98
gatifloxacin 164
gel electrophoresis 153
gelonin (rGel) 151, 180
gemcitabine 151
gemtuzumab ozogamicin (GO) 200
  for AML 200
  clinical utility 201–208
  de novo disease 207
  and *MDR1* gene 208
  in older patients 204
  pharmacology 201
  relapsed-refractory disease 206, 207
  second-generation trials 206, 207
  toxicity 205
gene splicing 103
glioblastoma 55, 137
  clinical trials 32–35
  IL-4 cytotoxin in 30
  IL-4R sites in 26
  IL-4-targeted therapy 23–35
  recurrent 23–35
  regression 34
glioblastoma multiforme (GBM) 25, 46, 47,
    58, 135, 142
  cells 101
  cpIL4-PE in 24
  diptheria-based immunotoxins
    97–108
  refractory and progressive 142
  standard chemotherapy 136

transferrin therapy outcome 145
  treatment of recurrent 24
glioma 137
  injection 33
  malignant 47, 58, 135
  recurrent 31, 57
  recurrent malignant 34
  specific cytotoxicity 53
  tumors 46
  in brainstem 146
glomeruli 107
glutathione 153
glutathione S-transferase 151
glycosyl phosphatidylinositol (GPI) 102
graft versus host disease (GVHD) 125, 129
granulocyte colony-stimulating factor
    (G-CSF) 163
granulocyte-macrophage colony stimulating
    factor (GMCSF) 128, 150
  receptor-targeted therapy 149–164
guanidine hydrochloride 153

hematologic malignancies 4, 5, 10, 11, 14,
    117, 129
  Pseudomonas exotoxins 4–14
hemolytic uremic syndrome (HUS) 12,
    80, 180
hairpin carboxylates 152
hairy cell leukemia (HCL) 4, 11
  BL22 activity 12
  cladribine-resistant 12
head and neck cancer 24
headache 123
hematemesis 190, 191
hematuria 12
hemoglobinuria 12
hemolysis 118
hemorrhage 178
hemorrhagic colitis 180
heparin-binding epidermal growth factor-like
    (HB-EGF) 152
hepatic
  encephalopathy 155
  necrosis 56
  toxicity 127, 205
herceptin 192
heterodimerization 101
high performance liquid chromatography
    (HPLC) 117, 191
high pressure liquid chromatography 153
high-dose chemotherapy (HDCT) 74, 83
histologic efficacy 57
histologically effective concentration
    (HEC) 57
Hodgkin/Reed-Sternberg (H-RS) 74